THE CHEMICAL EVOLUTION OF THE GALAXY

ASTROPHYSICS AND SPACE SCIENCE LIBRARY

VOLUME 253

THE CHEMICAL EVOLUTION OF THE GALAXY

by

FRANCESCA MATTEUCCI

Department of Astronomy,
University of Trieste

SPRINGER SCIENCE+BUSINESS MEDIA, B.V.

A C.I.P. Catalogue record for this book is available from the Library of Congress.

ISBN 978-1-4020-1652-3   ISBN 978-94-010-0967-6 (eBook)
DOI 10.1007/978-94-010-0967-6

Printed on acid-free paper

This book is dedicated to my son Marco

Contents

ABBREVIATIONS

AGB – Asymptotic Giant Branch

CHVC – Compact High Velocity Cloud

DLA – Damped Lyman-α Systems

GRB – Gamma Ray Burst

HB –Horizontal Branch

H-R – Hertzsprung-Russell

HVC – High Velocity Cloud

ICM – Intracluster Medium

IMF – Initial Mass Function

IR – Infall Rate

I.R.A.– Instantaneous Recycling Approximation

ISM – Interstellar Medium

IVC – Intermediate Velocity Cloud

K-H – Kelvin-Helmoltz

LSR – Local Standard of Rest

ML – Mass-Luminosity

MS – Main Sequence

MSTO – Main Sequence Turn-Off

NSE – Nuclear Statistic Equilibrium

PDMF – Present Day Mass Function

PNe – Planetary Nebulae

Pop I – Population I

Pop II – Population II

Pop III – Population III

QSO – Quasi Stellar Object

RG – Red Giant

RGB – Red Giant Branch

SFR – Star Formation Rate

S.M.L.A. – Sudden Mass Loss Approximation

SN I – Supernovae I

SN II – Supernovae II

SPH – Smooth Particle Hydrodynamics

SSP – Simple Stellar Population

VHVC – Very High Velocity Cloud

WD – White Dwarf

PREFACE

This book is based partly on a lecture course given at the University of Trieste, but mostly on my own research experience in the field of galactic chemical evolution.

The subject of galactic chemical evolution was started and developed by Beatrice Tinsley in the seventies and now is a flourishing subject. This book is dedicated to the chemical evolution of our Galaxy and aims at giving an up-to-date review of what we have learned since Tinsley's pioneering efforts. At the time of writing, in fact, books of this kind were not available with the exception of the excellent book by Bernard Pagel on "Nucleosynthesis and Chemical Evolution of Galaxies" (Cambridge University Press, 1997), and the subject of galactic chemical evolution has appeared only as short chapters in books devoted to other subjects. Therefore, I felt that a book of this kind could be useful.

The book summarizes the observational facts which allow us to reconstruct the chemical history of our Galaxy, in particular the abundances in stars and interstellar medium; in the last decade, a great deal of observational work, mostly abundance determinations in stars in the solar vicinity, has shed light on the production and distribution of chemical elements. Even more recently more abundance data have accumulated for external galaxies at both low and high redshift, thus providing precious information on the chemical evolution of different types of galaxies and on the early stages of galaxy evolution. The rapidly developing study of the distribution of gas along the galactic disk and the clouds of neutral hydrogen falling onto the disk, together with their chemical composition, has also provided another very important tool for understanding the formation of the galactic disk and spiral galaxies in general.

Following the review of the observational material, the book continues with a discussion of the basic ingredients necessary for constructing models of galactic chemical evolution: the stellar evolution and nucleosynthesis, the stellar birthrate, and the gas flows. Concerning stellar evolution and nucleosynthesis the book gives only short summaries of the basic facts necessary to calculate the element production from stars and the reader is directed to the many more detailed text books on the subject. A chapter is then devoted to the mathematical formalisms commonly used in galactic chemical evolution models either to compute analytical or numerical models. Analytical models do not allow a detailed calculation of the evolution of single chemical abundances but are very useful to understand some basic physical principles concerning chemical evolution. The main chapter of this book is perhaps where the formation and evolution of the Milky Way is inferred from the comparison of theoretical models with observational constraints. Finally, in the last chapter a comparison between the Milky Way and external spirals is presented.

My intention throughout the book has been to demonstrate how important can be the analysis of the chemical abundances and especially abundance ratios in galaxies for understanding their formation and evolution, and how, in spite of the many parameters involved in the modelling, we can draw important conclusions from a judicious comparison between models and observations. This last point, in my opinion, is particularly relevant in answering all those who are critical of modelling of galactic chemical evolution because of the many parameters involved.

The book is intended to be a review of the state of the art of the chemical evolution of the Galaxy and it contains many recipes for computing such evolution. In this sense it should be used as a cookbook by those who want to start working in this field and who already have some astronomical background. Since the subject of the book is in continuous development, I have included a fair number of references while trying at the same time to minimize them. For this reason, I apologize to those authors I have not quoted in the interest of brevity.

My hope is that the readers of this book, either students or experts, will find it useful and perhaps stimulating even to the point of personal participation in this field of research.

Francesca Matteucci

Chapter 1

OBSERVATIONAL EVIDENCE FOR
CHEMICAL EVOLUTION

1.1   OVERVIEW AND HISTORICAL PERSPECTIVE

Our Sun is located in a stellar system called the Milky Way galaxy. We know today that our Galaxy is similar to many other galaxies and these are the major structural units of the universe. In particular, we have obtained a considerable insight into the nature of our Galaxy from observations of external galaxies, which reveal large scale features that are hidden from us in the Milky Way.

As we look into the night sky we see an enormous number of stars with a large range of apparent brightnesses. On a clear night, we can also see a faint band of light, cut by a dark rift, stretching around the sky; the faint glow is the whole impression produced visually by our Galaxy, and the dark band is caused by obscuring dust. That is the Milky Way. In ancient Greece there was a legend saying that the Milky Way formed because a drop of milk was lost by goddess Hera while she was breastfeeding Heracles and the name *Galaxy*, which is the equivalent of *Milky Way*, originates from the Greek word *gala* which means milk. In ancient times the Milky Way has been considered in different ways in different cultures: for example, the Egyptians, the Teutons and the early Christians have all invoked divine origins or intent in the presence of our Milky Way. On the other hand, the Incas, with their pervasive interest in gold and the Polynesians living close to sea and its creatures seem to have drawn upon more immediate images from their own experience to interpret the celestial patterns.

The first scientific study of the physical nature of our Galaxy dates only from 1610, when Galileo observed the Milky Way and discovered that it could be resolved into *innumerable* faint stars. Therefore, it was realized that the diffuse light of the Milky Way could not any longer be attributed to some *celestial fluid* or represent the way to Valhalla or the road to Rome. The Milky Way became a stellar system.

By the middle of the eighteenth century, Thomas Wright and Immanuel Kant had offered a description of our Galaxy consisting of a disk of stars in which the

1

Sun is immersed. Kant remarked also that our Galaxy might not be the only one in the Universe and that similar systems, which he called *island universes*, might be distributed throughout space at large distances from our system.

Later, William Herschel, at the end of the eighteenth century, built telescopes large for that epoch, and with them studied the Galaxy and other stellar systems. His main achievement was to understand the shape of the Galaxy, which he concluded to be a flattened, roughly elliptical system with the Sun occupying the central position. He reached this conclusion by counting stars that he could observe to successive limits of apparent brightness in about 700 regions of the sky. He assumed that all stars have the same absolute brightness, that they are uniformly distributed in the sky, that the brightness falls as the inverse square of their distance and that he could observe until the border of the system. Then the ratios of the cube roots of the numbers of stars of a given brightness, seen in the field of his telescope in given directions, reflected the relative distances and hence he deduced the extent of the Galaxy. From this study he deduced that the Galaxy extends five times more in the plane of the Milky Way than in the direction perpendicular to it.

At end of the nineteenth century, the development of photography opened many new possibilities. Kapteyn studied some 200 Selected Areas distributed carefully over the sky and he started international collaborations to collect spectroscopic, proper motion and radial velocity data. Kapteyn and van Rhijn were able to estimate distances for stars of various apparent brightness and the distribution of stars in space. The problem of measuring distances was a fundamental one, since without knowing the distances to stars it was impossible to understand the extension of the Milky Way and decide either if the Galaxy included all the universe or if it had a limited spatial extent confirming the suggestion of Kant about the island universes. The method they used was the *proper-motion technique*; it consisted in measuring non-periodic motions of the stars across the sky and allowed the astronomers to extend the region of measurable distances beyond that obtained with the *parallax method*. The parallax method consists in measuring the shift in the apparent position of nearby stars due to the Earth's revolution around the Sun. This shift is inversely proportional to the distance to the star, so this technique is applicable only to nearby stars (usually less than 100 pc, 1 pc $\simeq 3 \cdot 10^{18}$ cm). They assumed again that the apparent brightness decreases as the square of the distance and that interstellar space is completely transparent. Now we know that this is incorrect, because there is strong absorption of starlight by interstellar material in the galactic plane unrecognized at that time. The picture that emerged from the Kapteyn study is known as the *Kapteyn Universe*: the Galaxy is a flattened spheroidal system of modest size, roughly five times more extended along the plane than in the vertical direction. This picture is similar to that of Herschel but Kapteyn added a scale to the system and a quantitative estimate of the density within it. The Kapteyn view was still almost an heliocentric one where the Sun was located slightly out of the galactic center at a distance of about 650 pc. He arrived at this conclusion from the fact that the star density was thought to decrease uniformly away from the center of the system. However, Kapteyn himself realized that there could also be an alternative explanation for his data for which the Sun was not near the center: if there was an absorbing interstellar medium then the stellar light would suffer extra dimming. If this dimming is incorrectly interpreted as a distance effect, then the distances of stars would be erroneous, leading to an artificial falloff of the star density in all directions away from the observer, so creating the impression of being at the center. He was however unable to adduce any evidence for such absorption.

Even before the Kapteyn model was published it was challenged by the results of Shapley in a series of papers published between 1915 and 1919. Shapley, at Mount Wilson Observatory, studied the globular clusters, which are compact spherical systems containing from 10^5 to 10^6 stars. Because of their great brightness these systems can be observed at very great distances from the Sun. In addition, since they lie above the galactic plane they do not suffer much interstellar absorption. Shapley pointed out that while the globular clusters are distributed uniformly above and below the plane, they are not uniformly distributed around the plane. That led him to conclude that, if these systems are major constituents of the Galaxy and are symmetrically distributed around the galactic center, then the Sun should not be in that center. He estimated that the radius of the distribution of globular clusters should be 100 kpc (1 kpc $=$ 10^3 pc) from the galactic center and the Sun should be at about 18 kpc from the galactic center. Shapley's conclusion that the Sun lies far away from the galactic center was indeed the correct one, as has been shown by all subsequent investigations, although today we estimate the distance of the Sun from the galactic center to be $\simeq 8.5$ kpc. However, by not considering the interstellar absorption Shapley was induced to believe that the Galaxy was a unique system, at least in the part of Universe accessible to observations. In fact, the dimensions of the Galaxy suggested by Shapley were 10 times larger than what Kapteyn had suggested and most of astronomers had believed. If the spiral nebulae external to our Galaxy, whose nature was still unrecognized, had the same dimensions as Shapley's Galaxy then they would have been inconceivably distant. For this reason many astronomers did not believe Shapley's conclusions. A famous debate on the size of the Galaxy and the nature of the spiral nebulae occurred during the annual meeting at the National Academy of Sciences held in Washington in April 1920 between Heber Curtis and Harlow Shapley.

The issue was then completely settled in the following years between 1920 and 1930 thanks to astronomers such as Edwin Hubble, Jan Oort and Bertil Lindblad. The major discovery of Oort and Lindblad was galactic rotation. They arrived simultaneously at the same conclusions by kinematical and dynamical considerations. Lindblad in 1926 had affirmed Shapley's location for the galactic center and developed a mathematical model for the rotation of our Galaxy. Against the Kapteyn Universe he argued that the mass of the Galaxy calculated by Kapteyn would produce a too weak gravitational field to retain the globular clusters. These objects, in fact, had been observed to have velocities of about 250 km sec^{-1} relative to the Sun, which is much larger than the escape velocity of the Kapteyn Universe. Lindblad also proposed that the system of globular clusters is nearly at rest relative to the Galaxy as a whole and that the Sun travels on a circular orbit at a velocity of the order of 200- 300 km sec^{-1}. Because a typical value for the velocity dispersion in stars of the solar region is 30 km sec^{-1}, he argued that all the low velocity stars had essentially the same energy of motion as the Sun and that they also traveled in circular orbits around the galactic center.

Oort arrived at the discovery of galactic rotation by studying high velocity stars and showed that they could be understood on the basis of Lindblad model. In particular, Oort discovered that below 62 km sec^{-1} radial velocities have a symmetric velocity distribution but above that critical velocity the directions of motion exhibit a marked asymmetry. This is what one should expect if the Sun belongs to a rapidly rotating system and moves on a circular orbit around the galactic center, while the high-velocity stars belong to a system which rotates much more slowly. In addition, Oort suggested that the galactic rotation is differential, namely the rotation velocity changes with the distance from the galactic center (that is, faster angular rotation near the center and slower near the edge).

In the meantime the American astronomer Edwin Hubble at Mount Wilson in 1923 was able to resolve the outer regions of two nearby spirals M31 and M33. He found Cepheid variables in these regions and applied the period-luminosity relation of Shapley (the one used to derive the distances to globular clusters). Hubble showed that the distances to the spiral nebulae were about 285,000 pc. These distances are large enough to prove that the spirals must be stellar systems as large as our own Galaxy.

So in 1927 the Kapteyn Universe had been superseded and our Galaxy had become part of the spiral nebulae, showing that Kant's intuition about the *island universes* was indeed correct. The reason that the star-counts had given such erroneous interpretations resides in the fact that interstellar absorption remained to be discovered by Trumpler.

In 1944 Walter Baade created a new insight into the study of galaxies when he resolved into stars the nucleus of M31, its two companions M32 and NGC205, and the elliptical galaxies NGC147 and NGC185. He realized that the brightest stars in these systems were completely different from the luminous blue stars found in the spiral arms. He then suggested that stars in a galaxy could be categorized into different *populations*. Baade described as *Population I* (Pop I) the blue stars associated with spiral arms and as *Population II* (Pop II) the red stars found in spheroidal components of galaxies such as bulges, halos and in globular clusters. The concept of stellar populations was very important because it triggered stellar evolution and star formation studies. Detailed spectroscopic analyses showed that Population I stars have metal abundances similar to the Sun, whereas Population II stars in the galactic halo and globular clusters are metal deficient by as much as a factor of 10^{-3} to 10^{-4} relative to the Sun.

In broadest terms, the Milky Way, as we know it now, is a spiral galaxy of morphological type Sbc (in the Hubble classification) and it can be described as a central flattened disk of gas and stars with a stellar bulge in the center, surrounded by a spherical halo of stars and globular clusters (see Figure 1.1). The Milky Way possesses also an extended and massive halo of dark matter (see Section 1.12.3) which should not be confused with the stellar halo to which we refer in the following simply as halo. The stellar populations of these main components (halo, bulge and disk) have quite different chemical compositions, kinematic and dynamical properties reflecting different evolutionary histories. The halo extends up to $\sim$ 30 kpc from the center, the bulge has an extension of $\sim$ 2 kpc and the disk is thin ($\simeq$ 200 pc thick) and extends up to $\sim$ 25 – 30 kpc from the center. The Sun is located at a distance of $R_\odot \sim 8.5$ kpc from the center.

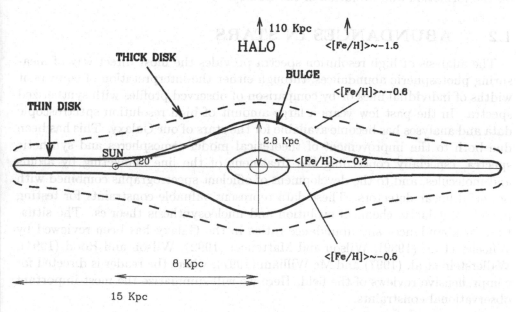

Figure 1.1. A sketch of the structure of the Milky Way: the stellar halo, bulge, thick and thin disk are indicated together with the mean metallicity of the stars in each galactic component. The galactocentric distance of the Sun together with the dimension of the optical disk are shown on the bottom. Reproduced here by kind permission of E. Zoccali and S. Ortolani.

The main observational evidence for the chemical evolution of the Milky Way and galaxies in general is provided by the fact that the stellar chemical composition is different in stars of different ages. In particular, there is evidence that the concentration of metals (all the elements heavier than He) is increasing with decreasing stellar age, indicating that there is a progressive metal enrichment of the interstellar medium (ISM) out of which stars form. The study of the chemical evolution of the Galaxy and galaxies in general is an attempt to reconstruct the history of the chemical composition of their gas through consideration of the processes of galaxy formation, star formation, stellar evolution, nucleosynthesis and possible gas flows. It is believed that all of deuterium, a major part of helium and some Li, were produced during the Big Bang, whereas the heavier elements are thought to be produced in stars. This is the simple scheme which we will adopt in studying the chemical evolution of the Galaxy.

In this chapter we will review the observational evidence for the chemical evolution of the Milky Way and therefore we will start by discussing the most important observational facts concerning abundances both in stars and in the ISM. Then we will review the observational information about the gas distribution along the galactic disk and the observed supernova frequency. Altogether these observational facts represent the most important observational constraints on the formation and evolution of the Galaxy.

1.2   ABUNDANCES IN STARS

The analysis of high resolution spectra provides the most direct way of measuring photospheric abundances, through either the interpretation of equivalent widths of individual lines or by comparison of observed profiles with synthetized spectra. In the past few years a large amount of high resolution spectroscopic data and analyses has become available for the stars of our Galaxy. This has been due both to the improvement of theoretical model atmospheres and synthetic spectra, especially concerning the treatment of the line blanketing by atoms and molecules, and to the development of efficient spectrographs combined with powerful linear detectors. These data represent valuable constraints for testing models of galactic chemical evolution and nucleosynthesis theories. The situation for abundances and abundance ratios in the Galaxy has been reviewed by Wheeler et al. (1989), Wilson and Matteucci (1992), Wilson and Rood (1994), Wallerstein et al. (1997) and Mc William (1997), where the reader is directed for comprehensive reviews of the field. Here we will summarize the most important observational constraints.

Before discussing the abundances in stars we should define the "bracket" notation of abundance workers. If A represents some measured abundance by number relative to hydrogen in a program star and in a standard star, which is the Sun, then:

$$[A] = log(A)_{programstar} - log(A)_{Sun} \qquad (1.1)$$

The "bracket" abundances are measured in units of *decimal exponentials* or *dex*; for example, if two stars differ by 1 dex in their iron abundance, it means the abundances differ by a factor of 10.

Normally, [Fe/H], the iron abundance relative to hydrogen abundance, is used to indicate the metallicity of a star but in reality it gives only the abundance of iron, which is the most commonly derived abundance in late-type stars. On the other hand, when discussing theoretical models the so-called *metallicity* is the global metal abundance including all the elements heavier than He. It is very important to distinguish between global metallicity and iron abundance since, as we will see in the next chapters, oxygen (which dominates the global metallicity) and iron have abundances evolving in a very different way.

Normally, the reference abundances are the solar abundances, both photo-spheric and meteoritic, which are shown in Table 1.1. Photospheric absorption lines (Fraunhofer lines) provide the number ratios X/H for the most common chemical species with suitable features in the optical or infrared from neutral and singly ionized ions or molecules such as CH, CN, CO, OH, MgH. The photospheric analysis shows that the number ratios of some of the most common elements, $\frac{(C+N+O)}{H}$ is close to 10^{-3} and that of heavier metals (Na, Mg, Al, Si, Ca and Fe) is 10^{-4}. The abundances of common elements such as He, Ne and Ar cannot be determined in this way in late-type stars and most information comes from hot stars or nearby HII regions like Orion. Concerning the light elements, D is only destroyed inside stars and therefore no or little D is found in the solar or other photospheres; what is measured is always a lower limit to the primordial abundance. Li is also partly destroyed in stars and is found to be reduced by two orders of magnitude in the Sun while Be and B may have suffered some destruction as well, but only by factors of the order of 2 which are inside the observational uncertainties. Finally, there is limited information on rare elements such as U and Th, rare earth elements and on isotopic ratios which are restricted to those which can be determined from molecules, e.g. C and Mg.

If we define X, Y, Z as the mass fractions of hydrogen, helium and metals, then the primordial chemical composition is X~0.76 , Y~0.24 and Z=0.00 while the *standard solar mass fractions* we will refer to since now on are: X~0.70, Y~0.28 and Z~0.02.

Meteoritic abundances come from three main types: stony, stony irons and irons. They probably result from the break up of small parent bodies (asteroids) which were not subjected to the kind of differentiation that affects major planets like the Earth. Generally, the agreement between meteoritic and photospheric abundances is now quite good with the exception of the particular cases of Li, Be and B. It was generally believed that meteoritic abundances are more reliable than photospheric ones, these latter being subjected to the uncertainties in the

Table 1.1. Solar abundances relative to $logN_H = 12.0$; meteoritic abundances relative to $logN_{Si} = 7.55$. Data are from Anders and Grevesse (1989) updated by Grevesse, Noels and Sauval (1996). Values in brackets are based on solar wind and energetic particles corrected for first ionization potential effect. When two decimal places are given, the error estimate is ±0.10 or better. Table from Pagel (1997).

Chemical element	Photospheric	Meteoritic	Chemical element	Photospheric	Meteoritic
H	12.00		Sc	3.18	3.09
He	(10.99)		Ti	5.03	4.93
Li	1.16	3.30	V	4.00	4.01
Be	1.15	1.41	Cr	5.67	5.68
B	2.6:	2.78	Mn	5.39	5.52
C	8.55		Fe	7.50	7.49
N	7.97		Co	4.92	4.90
O	8.87		Ni	6.25	6.24
F	4.6:	4.47	Cu	4.21	4.28
Ne	(8.08)		Zn	4.60	4.66
Na	6.33	6.31	Rb	2.60	2.40
Mg	7.58	7.57	Sr	2.97	2.91
Al	6.47	6.48	Y	2.24	2.22
Si	7.55	7.55	Zr	2.60	2.60
P	5.45	5.52	Ba	2.13	2.21
S	7.30	7.19	La	1.17	1.21
Cl	5.5:	5.27	Ce	1.58	1.62
Ar	(6.5)		Nd	1.50	1.48
K	5.1	5.12	Eu	0.51	0.54
Ca	6.36	6.34	Th		0.08

oscillator strengths. In any case the disagreement between photospheric and meteoritic determination for most of the elements are within 0.2 dex.

Generally, the solar abundances are referred to as *cosmic abundances* implying that the solar chemical composition represents the universal composition of matter. This is not correct since we now know that there are abundance and abundance ratio variations from star to star and from galaxy to galaxy; therefore, it is preferable to refer to these abundances as *local abundances*.

1.2.1   STELLAR SYSTEMS IN THE GALAXY

As already mentioned, Baade first introduced the concept of stellar populations and since then it is generally adopted the view of dividing the stars in the Galaxy into two main populations: Pop I and Pop II distinguished by spatial distribution, age, kinematics and chemical composition. Pop I was identified with the disk population containing young, metal rich stars. Their orbits are nearly circular, close to the galactic plane and follow the general rotation of the disk. This means that they possess low velocity dispersion and low velocities relative to the Sun. Pop II was instead identified with the spheroidal system or halo, sparse in the solar vicinity but more concentrated to the center of the Galaxy. This population is traced by globular clusters and high velocity stars with elongated orbits. All of these stars are old and metal poor (by factors up to 1000 less than the Sun).

At the present time, the situation is far more complicated than that originally described by Baade; in fact, we can envisage at least four distinct stellar populations: *halo population* which can be identified with the Pop II of Baade, *bulge population* which shares all of the kinematical properties of the halo population (see Figure 6.11) but the dominant metallicity of the stars is solar and possibly even larger, *thin-disk population* which can be identified with the Pop I of Baade and *thick-disk population* which represents a population intermediate between the halo and thin-disk as far as the kinematical and chemical properties are concerned.

In summary, the properties of the main stellar populations are:

- *halo population*: stars with metallicities [Fe/H] < −1.0 dex, height above the galactic plane larger than 920 pc and eccentric and elongated orbits. They are the oldest stars in the Galaxy.

- *bulge population*: stars with metallicities in the range $-1.5 < [Fe/H] < +1.0$ and kinematics like the halo stellar population. They are also old stars as the halo stars.

- *thick-disk population*: stars with metallicities $-1.0 \leq [Fe/H] \leq -0.6$ dex, height above the galactic plane between 270 and 920 pc and kinematics intermediate between the halo and thin disk stars. They are probably younger than the halo stars unless they have been accreted from external satellites.

- *thin-disk population*: stars with metallicities [Fe/H] > −0.6 dex, height above the galactic plane inside 270 pc and circular orbits. Their ages span a large range from ∼ 9-10 Gyr to few million years.

1.2.2 THE MOST METAL POOR STARS IN THE GALAXY

Most of the data we are discussing refer to the *solar neighbourhood* which is generally defined as the cylindrical region, centered on the Sun with 1 kpc radius and 1 kpc height. The most metal poor stars in the Galaxy belong to the halo and they are very important to analyze since their chemical composition reflects the composition of the ISM in the early stages of Galaxy evolution. This in turn, represents a useful tool for understanding the high-redshift universe.

In general, it is found that very metal-poor and metal-poor stars ([Fe/H]<-1.0 dex), belonging to the halo population, show a marked overabundance of α-elements (namely all the elements built from α-particles such as O, Ne, Mg, Si, S, Ca and Ti) relative to iron, in other words [α/Fe] >0 (see Figures 1.2 and 1.3). The meaning of such an overabundance in terms of galactic chemical evolution will appear more clearly at the end of this book. At the moment we only describe what is observed without any interpretation. As we will see in the next chapters, overabundances of α-elements relative to iron are particularly important in order to understand the history of the nucleosynthesis and the star formation in our Galaxy.

Many studies of very metal poor stars have been performed but we mention here only the most recent ones at the time of writing. These last studies are from McWilliam et al. (1995) and Ryan et al. (1996). In the first study, the abundances of several heavy elements are derived for 33 extremely metal deficient stars (mainly subdwarfs), with $-4.0 \leq [Fe/H] \leq -2.0$ dex.

Concerning the α-elements, they found [Mg/Fe], [Si/Fe] and [Ca/Fe] to be overabundant relative to the solar value, extending and confirming previous studies of more metal rich halo stars. In particular, they found Mg, Si and Ca to have roughly the same average overabundance ($< [\alpha/Fe] >=+0.44 \pm 0.02$ dex) whereas [Ti/Fe]=+0.31 dex. However, [Mg/Fe] at very low metallicities seems to decrease whereas [Ti/Mg] and [Ca/Mg] increase by $\simeq 0.2$ dex from [Fe/H]=-2.0 to -4.0 dex. They also revealed previously unnoticed trends of [Cr/Fe], [Mn/Fe] and [Co/Fe] with [Fe/H]; both [Cr/Fe] and [Mn/Fe] show a decline of about 0.5 dex with decreasing iron, between [Fe/H]=-2.4 and -4.0 dex (see Figure 1.4). On the other hand, [Co/Fe] increases by about 0.5 dex as the iron declines from -2.4 to -4.0 dex. They confirmed the well known decline in [Al/Fe], [Sr/Fe] and [Ba/Fe] with decreasing [Fe/H] (see Figure 1.5). The remarkable fact is that all of these six abundance ratios show a sudden change of slope near [Fe/H]=-2.4 dex. They concluded that a distinct phase of nucleosynthesis occurred before the Galaxy reached [Fe/H]=-2.4 dex, their preferred scenario being that of variable stellar yields (namely the amounts of different chemical elements produced and ejected into the ISM by stars, see chapter 2) changing with either mass or metallicity or both.

Finally, another remarkable result of this study are the large individual deviations (up to 3 dex for a given [Fe/H]) observed in the abundance ratios of heavy elements, especially [Sr/Fe] and [Ba/Fe] (these elements are *s-process* elements, see chapter 2), which they interpreted as intrinsic and due to an inhomogeneous early Galaxy. The same conclusion had been already reached by Primas et al. (1994) from the spread they had found in [Sr,Ba/Fe] ratios at very low metallicities.

Ryan et al. (1996) presented an abundance analysis for 19 metal-poor subdwarf stars, for elements between Mg and Eu. All the stars they analyzed have [Fe/H]<

−2.5 dex. These authors concluded that the α-elements Mg, Si, Ca and Ti possess almost uniform overabundances (relative to iron) down to at least [Fe/H] =-4.0 dex; [Mg/Fe] increases slightly for [Fe/H]< −2.5 dex but the slope is only -0.15 dex per dex which they claim may be due to systematic errors. Stars with [Fe/H]< −2.5 dex reach a plateau in [Al/Fe]∼ −0.8 dex which extends at least down to [Fe/H] =-4.0 dex. They confirmed the underabundance of [Cr/Fe] and [Mn/Fe], and the overabundance of [Co/Fe] in stars with [Fe/H] < −2.5 dex. They also found a mild overabundance of [Ni/Fe] although not so pronounced as that of Co (see Figure 1.4). Concerning the s-process elements Ba, Sr and Y, they showed that [Sr/Fe] has a spread larger than 2 dex at [Fe/H] < −3.0 dex, greatly exceeding reasonable errors in the measurements and analyses. [Ba/Fe] decreases for [Fe/H] < −2.0 dex and exhibits less scatter than [Sr/Fe].

In Figures 1.2-1.3 are assembled most of the existing data relative to C, N, O, Mg, Si, S, Ca and Fe for solar neighbourhood stars of all metallicities at the time of writing. This sample was obtained by selecting the most recent and higher quality abundance data from many sources and renormalizing them to the same solar abundances (meteoritic values). These data have been analyzed with a statistical method which allows one to quantify the observational spread in measured elemental abundances. After applying this statistical method one obtains three lines which summarize the trend of the data as a function of [Fe/H] (summary lines): the midmean -an average of all data between quartiles of the distribution of relative abundances over a given range of [Fe/H]; the lower semi-midmean -the midmean of all observations below the median; the upper semi-midmean - the midmean of all the observations above the median.

There is evidence for a slight increase of the [O/Fe] ratio for [Fe/H]< −1.0 dex whereas the other α-elements, with the exception of Mg for which a slight trend is also observed, show a flatter behaviour (plateau) for [Fe/H]< −1.0 dex. In Figures 1.4 and 1.5 the abundances of the Fe-peak elements and the s-process elements relative to iron are shown.

Finally, Christlieb et al. (2002) reported the discovery of the most metal poor star ever observed: its Fe abundance is [Fe/H]=-5.3 dex.

1.2.3 THE DISK POPULATION

The most important relatively recent study on disk stars is from Edvardsson et al. (1993) who derived abundances of O, Na, Mg, Al, Si, Ca, Ti, Fe, Ni, Y, Zr, Ba and Nd as well as individual photometric ages for 189 nearby field F and G disk dwarfs. These data are included in Figures 1.2-1.3. They also estimated, from kinematic data, the orbital parameters of the stars and derived the distances from the galactic center of the star birthplaces. Their relative iron abundances and abundance ratios relative to iron are estimated, for most elements, to have an accuracy with a standard deviation of 0.05 dex.

12

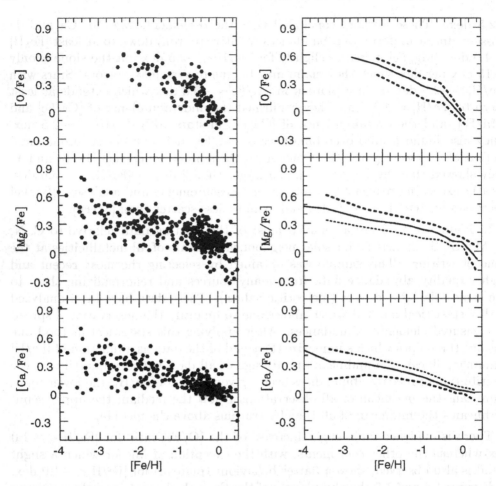

Figure 1.2. Abundance ratios of α-elements (O, Mg, Ca) to iron ([α/Fe]) as functions of the metallicity [Fe/H]. The left panels show the data from various sources normalized to the same solar abundances (Anders and Grevesse 1989, meteoritic abundances), whereas the right panels show the summary lines (the solid line represents the midmean and the dotted lines represent the lower and upper semi-midmean, respectively), as described in the text. Reproduced by kind permission of C. Chiappini.

Metal poor ([Fe/H] < −0.4 dex) disk stars are shown to be relatively over-abundant in the α-elements confirming all the previous studies, while a rather interesting finding is that [α/Fe] ratios for these metal poor stars seem to decrease slightly with increasing galactocentric distance. A similar trend seems to be observed also in halo stars; Nissen and Schuster (1997) identified some metal rich halo stars ([Fe/H] $\sim$ -1.0 dex) with both high and low [α/Fe] ratios. They found that the halo stars with low [α/Fe] ratios tend to be on orbits biased to the outer halo, thus perhaps indicating a decrease of these ratios with galactocentric

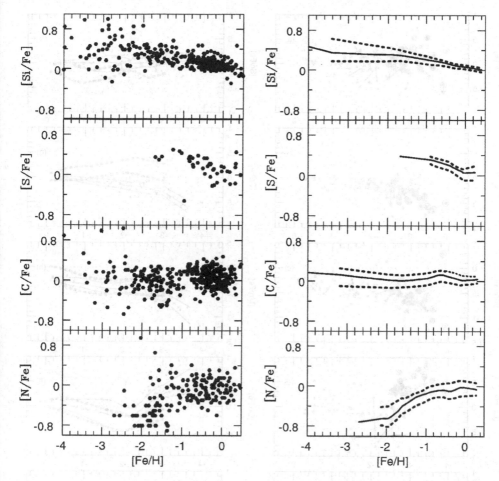

Figure 1.3. The same as Figure 1.2 for Si, S, C and N. Reproduced by kind permission of C. Chiappini.

distance. As we will elaborate later, this trend, if real, can be suggestive for the mechanism of Galaxy formation.

Also an interesting fact is that the scatter found in abundance ratios such as [Si/Fe] is only about 0.05 dex or less, which is about 54 times less than the corresponding scatter in [Fe/H] versus stellar age (see also Section 1.6).

1.2.4 ELEMENT BY ELEMENT

We discuss some elements in particular:

- *Oxygen* deserves a special discussion: the main concern in abundance studies is with the paucity of samples of significant sizes analyzed in a homogeneous

14

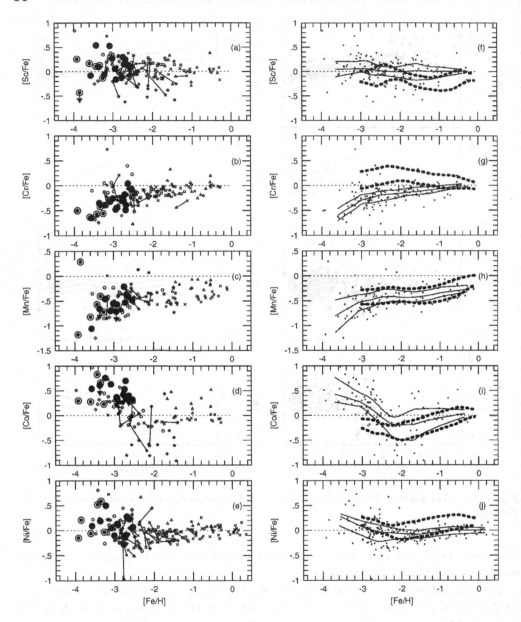

Figure 1.4. Abundances of iron peak-elements (Sc, Cr, Mn, Co, Ni) in the left panels. Multiple observations of a star are shown by solid lines. In the right panels are shown the summary lines (continuous curves) as defined for the data in Figures 1.2 and 1.3, and the model predictions of Timmes et al. (1995) (dashed curves) superimposed on the data. From Ryan et al. 1996, Ap.J. Vol. 471, 254; reproduced here by kind permission of S. G. Ryan and the University of Chicago Press (copy right 1996).

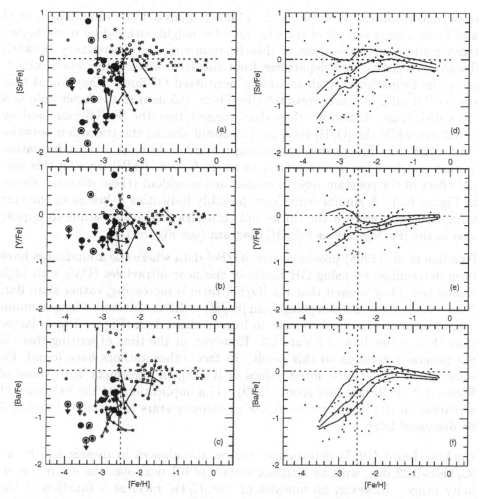

Figure 1.5. The same as Figure 1.4 for the abundances of neutron-capture elements (Sr, Y, Ba). In the right panel are shown only the summary lines. From Ryan et al. 1996, Ap.J. Vol. 471, 254; reproduced here by kind permission of S.G. Ryan and the University of Chicago Press (copy right 1996).

way, the possible existence of systematic offsets between different sets of data and the discrepancy between O abundances determined using high excitation permitted and low excitation forbidden lines. The most reliable oxygen abundance data are probably those derived from the forbidden line λ 6300, which unfortunately becomes too weak to measure in the most metal poor stars, particularly in dwarfs, and consequently one has to rely on the stronger permitted lines, which are highly sensitive to temperature and non-LTE (local thermodynamic equilibrium) effects, or on OH bands which also involve some uncertainties. Gratton et al. (1997a) analyzed the abundances of O and Fe in a large sample of stars in the solar neighbourhood by using higher temperatures in the analysis of dwarfs, removing the discrepancy between dwarfs and giants, and departures from the LTE assumption when considering the formation of high excitation permitted OI lines. They found that the [Fe/O] ratio is nearly constant ([Fe/O] $\simeq$ -0.5 dex) in the inner halo and thick-disk stars. Moreover, their data suggest that the [Fe/O] increased by $\simeq$ 0.2 dex while the [O/H] ratio held constant during the transition between thick- and thin-disk phases, suggesting a sudden decrease in star formation at that epoch. They plotted the data as [Fe/O] vs. [O/H] since in this way the effect of the constant oxygen abundance is evident (these data are shown in Figure 6.5). A similar behaviour, possibly indicating a hiatus in the star formation rate between the thick- and thin-disk formation, seems to appear also in the [Fe/Mg] versus [Mg/H] diagram (see Figure 1.9).

Israelian et al. (1998) presented new [O/Fe] data where the abundances have been determined by using OH bands in the near ultraviolet (UV) with high resolution. They showed that the [O/Fe] ratio is increasing, rather than flat, from +0.6 dex to +1.0 dex going from [Fe/H] =-1.5 to -3.0 dex. The maximum overabundance of oxygen relative to iron in these data (Figure 1.6) is larger than that in the data in Figure 1.2. However, at the time of writing there is not general consensus on this result. In fact, other authors have found, for some of the same stars, lower values of [O/Fe], in agreement with those of Figure 1.2 (Fulbright and Kraft, 1999). The implications of the existence of a plateau in the [α/Fe] ratios in low metallicity stars are manifold and will be discussed later.

- *Carbon:* Laird (1985) determined carbon abundances in dwarfs and found [C/Fe]=-0.22 dex, with an intrinsic scatter of 0.1 dex, over the entire metallicity range. However, an analysis of the [C/Fe] ratio as a function of the effective temperature indicated a systematic offset, so a correction of 0.20 dex was applied to all the [C/Fe], thus giving roughly solar values. Tomkin et al. (1986) examined 32 halo dwarfs with [Fe/H] in the range -2.5 to -0.7 dex and found that [C/Fe]= -0.2 $\pm$0.15 dex for [Fe/H]> -1.8 dex. Carbon et al. (1987) derived C abundances for 83 dwarfs in the range -2.5 $\leq [Fe/H] \leq -0.6$ dex. They found a solar [C/Fe] ratio over most of the metallicity range, with a star to star scatter of 0.18 dex. They also noticed an increase of [C/Fe] at the very low metallicities. However, this upturn is sensitive to the assumed O abundance. Wheeler et al. (1989) reanalyzed all the three surveys and attempted to place them onto a common effective temperature scale but the overall trend of increasing [C/Fe] ratio at low metallicities remained.

Andersson and Edvardsson (1994) determined the [C/Fe] ratio in 85 dwarfs in the metallicity range $-1.0 \leq [Fe/H] \leq +0.25$ dex and found that this

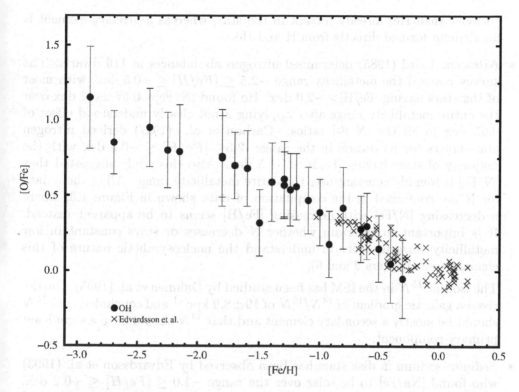

Figure 1.6. The [O/Fe] ratio as a function of [Fe/H] for a data sample observed by Israelian et al. (1998) (full circles) and by Edvardsson et al. (1993) (crosses). The full circles represent abundances derived from OH lines whereas the crosses indicate abundances derived from the IR OI triplet. These latter have been shifted by +0.05 dex in order to reproduce the same abundance scale as in Israelian et al. (1998). From Israelian et al., 1998, Ap.J. Vol. 507, 805; reproduced here by kind permission of G. Israelian and the University of Chicago Press (copy right 1998).

ratio is slowly decreasing with time and increasing metallicity in the disk. Moreover, Tomkin et al. (1995) determined C abundances in 105 dwarfs in the range $-0.8 \leq [Fe/H] \leq +0.2$ dex and found a moderate enrichment of C in metal-deficient stars, such as [C/Fe]=+0.2 ± 0.05 dex at [Fe/H]=-0.8 dex. Such a behaviour is qualitatively similar to that seen for α-elements and oxygen. All of these data are contained in the compilation of data shown in Figure 1.3.

The carbon isotope ^{13}C seems to show a predominantly secondary origin confirmed by the behavior of the $^{12}C/^{13}C$ ratio in the ISM along the galactic disk where it increases with decreasing galactocentric distance. A *secondary* element from the nucleosynthetic point of view, as we will see better in chapter 2, is an element formed in proportion to the abundance of heavy elements (heavier than He) already present in the star, whereas a *primary* element is an element formed directly from H and He.

- *Nitrogen*: Laird (1985) determined nitrogen abundances in 116 dwarfs. The survey covered the metallicity range $-2.5 \leq [Fe/H] \leq +0.5$ dex, with most of the stars having [Fe/H]> -2.0 dex. He found [N/Fe]$=-0.67 \pm 0.2$ dex over the entire metallicity range after applying a not clearly understood offset of 0.65 dex to all the [N/Fe] ratios. Carbon et al. (1987) derived nitrogen abundances for 83 dwarfs in the range -2.5$\leq [Fe/H] \leq -0.6$ dex with the majority of stars having [Fe/H] < -1.5 dex. Also this study suggested that [N/Fe] is roughly constant over the entire metallicity range. All of these data for N are contained in the compilation of data shown in Figure 1.3, where a decreasing [N/Fe] with decreasing [Fe/H] seems to be apparent instead. It is important to ascertain whether N decreases or stays constant in low metallicity stars in order to understand the nucleosynthetic nature of this element (see chapters 2 and 6).

  The isotope ^{15}N in the ISM has been studied by Dahmen et al. (1995) who derived a galactic gradient of $^{14}N/^{15}N$ of 19 ± 8.9 kpc^{-1} and concluded that ^{15}N should be mostly a secondary element and that ^{14}N should have a significant primary component.

- *Sodium*: sodium in disk stars has been observed by Edvardsson et al. (1993) who found [Na/Fe] to be solar over the range $-1.0 \leq [Fe/H] \leq +0.2$ dex. This behaviour, if compared with that of [Mg/Fe] suggests that [Na/Mg] shows a clear odd-even effect.

- *Aluminium*: considering Al in disk stars, Edvardsson et al. suggested that [Al/Fe] mimics the rise of [Mg/Fe] with decreasing [Fe/H] but at a reduced level. As a consequence, the [Al/Mg] ratio barely shows any odd-even effect over the metallicity range of this study. ^{26}Al is a radioactive element decaying to ^{26}Mg (half-life $1.1 \cdot 10^6$ years). This decay gives rise to the 1809 and 1130 keV gamma-ray lines. Timmes et al. (1995) suggested that the $2M_\odot$ of ^{26}Al, derived from diffuse emission of the above lines, can be explained by the ^{26}Al injected into the ISM in the last million years by massive stars.

- *Potassium*: potassium so far has been measured only by Gratton and Sneden (1987). They examined 23 stars including dwarfs and giants, in the metallicity range $-2.3 \leq [Fe/H] \leq +0.3$ dex. Their results suggested that [K/Fe]≥ 0 dex in metal poor stars, but with a very large scatter.

- *Cu and Zn*: the abundance of the first nuclei beyond the iron peak, such as Cu and Zn, have been measured both for halo and disk stars (Sneden et al. 1991 and references therein). Copper relative to iron shows an increase with [Fe/H] whereas [Zn/Fe] is solar over all the studied metallicity range. In

Figures 1.7 and 1.8 are shown the data for [Cu/Fe] and [Zn/Fe] as functions of [Fe/H], respectively. The study of the evolution of the abundance of Zn has become particularly interesting recently since it is observed in high redshift objects such as Damped Lyman-α systems (DLA). Among the different classes of quasar absorbers, the DLA systems are those characterized by a column density $N(HI) \geq 10^{20} cm^{-2}$ and by many low ionization lines such as FeII, SiII, CrII, ZnII and OI. These objects are unknown and they could either represent the early phases of the evolution of galactic disks or they could be similar to the local extragalactic HII regions (small irregular gas rich galaxies). A detailed analysis of their abundances and abundance ratios represent at the moment the only way of inferring their nature.

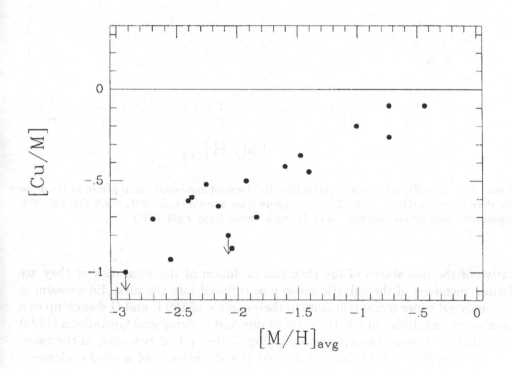

Figure 1.7. The [Cu/M] versus [M/H] = [Fe/H]. The horizontal solid line is drawn at [Cu/M] = 0. From Sneden et al. 1991, A&A Vol. 246, 354; reproduced here by kind permission of Springer Verlag (copy right 1991).

1.2.5   SUPER METAL-RICH STARS

About 4% of local disk stars are super metal-rich stars (defined as more metal rich than the Hyades with [Fe/H]= +0.12 dex). These stars are either represen-

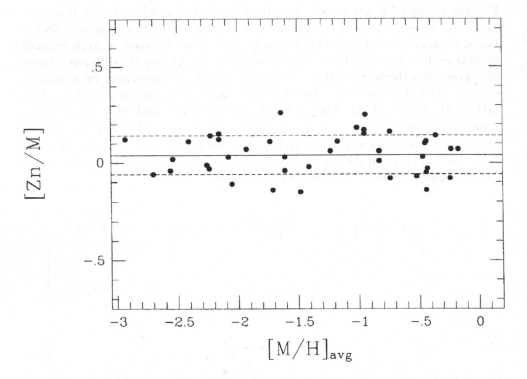

Figure 1.8. The [Zn/M] versus [M/H] = [Fe/H]. The solid horizontal line is placed at the mean Zn abundance [Zn/M] = +0.04. Data and figure from Sneden et al. 1991, A&A Vol. 246, 354; reproduced here by kind permission of Springer Verlag (copy right 1991).

tative of the last stages of the chemical evolution of the local disk or they are former members of the galactic bulge now diffused into the disk. Edvardsson et al. analyzed super metal rich stars in their sample of 189 F and G dwarfs up to a maximum metallicity of [Fe/H] $\sim$ +0.26 dex and Feltzing and Gustafsson (1998) extended the previous sample by analyzing 47 disk metal-rich stars in the range $-0.1 < [Fe/H] < +0.42$ dex and derived the abundances of several α-elements and Fe and, for the first time in such metal rich stars, the abundances of Cr, Mn and Co. They found that the abundances of Mg, Al, Si, Ti, Ca and Cr follow that of iron whereas the abundance of oxygen seems to decline with increasing [Fe/H]. Castro et al. (1997) have presented a high resolution abundance analysis of 9 super-metal-rich high velocity dwarfs in the solar vicinity (5 of them have [Fe/H] $\geq$ +0.4 dex). The [α/Fe] ratios in these stars are found to decrease with increasing metallicity thus extending the trend observed for the less metal rich disk stars, and the [s-process/Fe] ratios are underabundant by $\approx$ -0.3 dex. It is very important to have an indication of abundance ratios in super-metal-rich

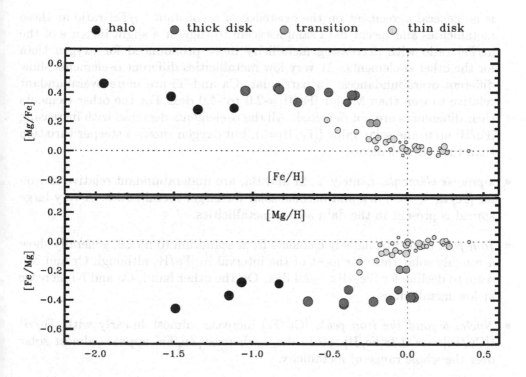

Figure 1.9. Top: [Mg/Fe] versus [Fe/H] in the solar neighbourhood. Circle diameters indicates age estimates, with small diameters indicating the youngest stars. Different stellar populations are given with various greyscale symbols as indicated in the legend on the top. Bottom: [Fe/Mg] versus [Mg/H]. The abrupt change in [Fe/Mg] ratios, is indicative of an intermediate phase between the thick- and the thin-disk characterized by a substantial release of Fe and otherwise low or no Mg production. Compare with Figure 6.5. From Furhmann 1998, A & A Vol. 338, 161; reproduced here by kind permission of Springer Verlag (copy right 1998).

stars both to improve our knowledge of galactic evolution and also to establish if these stars belong to the disk or to the bulge, as will be seen in the next chapters.

1.2.6 SUMMARY OF FIELD STARS ABUNDANCES

- *Carbon* is observed to vary in lockstep with iron ([C/Fe]=[Fe/H]) although a slightly increasing trend with decreasing metallicity may be present.

- *Nitrogen* seems to vary in lockstep with iron for most of the metallicity range, although a decrease at low metallicities cannot be excluded. A large scatter is present in the data.

- *α-elements*, there is a general consensus about these elements being over-abundant relative to iron in halo stars from [Fe/H]=-1.0 to -4.0 dex. There is no general agreement on the existence of a constant [α/Fe] ratio at these metallicities and recent data samples seem to suggest a slight increase of the [α/Fe] ratio with decreasing metallicity more pronounced for oxygen than for the other α-elements. At very low metallicities different α-elements show different overabundances: in particular, Ca and Ti are more overabundant relative to iron than Mg for [Fe/H]=-2.0 to -4.0 dex. For the other elements clear differences are not detected. All the α-elements decrease with increasing [Fe/H] up to the solar value ([Fe/H]=0), but oxygen shows a steeper variation than the other elements.

- *s-process elements*, namely Y, Sr and Ba, are underabundant relative to iron for [Fe/H] < −2.5 dex and become solar for larger metallicities. A very large spread is present in the data at low metallicities.

- *Iron-peak elements*, namely elements from scandium to nickel, generally show a roughly solar ratio for most of the interval in [Fe/H] although Cr and Mn seem to decline for [Fe/H]< −2.4 dex. On the other hand, Co and Ni increase at low metallicities.

- *Nuclei beyond the iron peak*, [Cu/Fe] increases almost linearly with [Fe/H] ([Cu/Fe]= $\simeq$ 0.38[Fe/H] +0.15 dex), whereas [Zn/Fe] appears almost solar over the whole range of metallicity.

1.3 ABUNDANCES IN GLOBULAR CLUSTERS

Observational and theoretical evidence has long established that galactic globular clusters contain the oldest stars in the Galaxy. They also represent a unique opportunity to study the properties of a simple stellar population (SSP), namely a population of stars that are coeval and have the same chemical composition (with the exception of ωCen and $M92$). Therefore, the measure of the abundances in globular cluster stars can represent a unique test for reconstructing the early dynamical and chemical evolution of the Galaxy. The metal abundance of globular clusters is traditionally indicated by the [Fe/H] ratio, which is generally assumed as representative of the total metal abundance. This is not correct, since the iron abundance does not evolve in lockstep with oxygen, which is the dominant element in the global metallicity Z, as shown by the observational data discussed above. Unfortunately, this important fact is often overlooked in the literature and sometimes theoretical models, devised to calculate the global metallicity, are compared to data referring to iron. Moreover, many observers indicate with [Fe/H] any heavy element abundance, including those of the α-elements.

One of the difficulties in measuring abundances in globular clusters was that these stars are generally very faint, and therefore it is difficult to reach a high resolution and signal-to-noise ratio. In the last years, the situation has noticeably improved and the most recent and accurate determination of [Fe/H], based on high dispersion spectroscopy of red giants, can be found in Carretta and Gratton (1997). This study confirms the well established earlier work that globular clusters are metal poor relative to the Sun, their [Fe/H] ranging from ~ -0.64 dex down to ~ -2.12 dex.

From the point of view of the study of the formation and evolution of the Galaxy, to be discussed in chapter 6, it is very important to obtain, together with the abundance of iron, the abundances of the α-elements (O, Mg, Si, Ca) and compare the derived trends with those observed in field stars. In particular, it is interesting to see if there is an overabundance of α-elements relative to iron, as observed in metal poor field stars. However, the abundance of oxygen, together with those of C and N, should be interpreted with care, because they can show a significant depletion due to deep mixing of the stellar envelope through the CNO hydrogen burning shell where the oxygen is transformed into ^{14}N (see chapter 2). Most of the high resolution data whether obtained in past or in recent years seem to indicate the existence of an overabundance of α-elements (Mg, Si, Ti) relative to iron in globular clusters, with this excess seeming to change slightly from cluster to cluster. In Figure 1.10 we show the average $< [\alpha/Fe] >$ in globular clusters of different metallicities. From this figure one can see that there is no variation in $< [\alpha/Fe] >$ as a function of either metallicity or age. This may indicate that each cluster was polluted mainly by massive stars, which produce an above-solar $[\alpha/Fe]$ ratio (see chapter 2). Therefore, a comparison of the trend $[\alpha/Fe]$ versus $[Fe/H]$ for halo field stars with the $< [\alpha/Fe] >$ ratio of globular clusters is not appropriate, since for halo stars there has been a *progressive* chemical enrichment which does not exist in globular clusters (homogeneity in the metal content). There is, however, an exception which is represented by two clusters (Pal 12 and Rup 106) which are known as young clusters and show lower overabundances of α-elements relative to iron. Since these clusters have space velocities and positions consistent with membership of the Magellanic Stream, they may have been accreted.

Concerning the abundances of neutron capture elements in globular clusters stars, Armoski et al. (1994) concluded that the abundances of the elements considered (Ce, Nd, Ba, Eu) are virtually indistinguishable from those determined for field stars of similar metallicity.

In discussing the derivation of the abundances in globular clusters, one should mention, in addition to the high resolution spectroscopy, the so-called *metallicity indicators*. Here we will briefly summarize the most common metallicity indicators in globular clusters :

- *color-magnitude diagram indices*; detailed studies of the color-magnitude diagram of globular clusters have allowed the definition of several metallicity indices. Most of them are based on the sensitivity of the color of the red giant branch (RGB) to the global metallicity. Among those indices we recall:

24

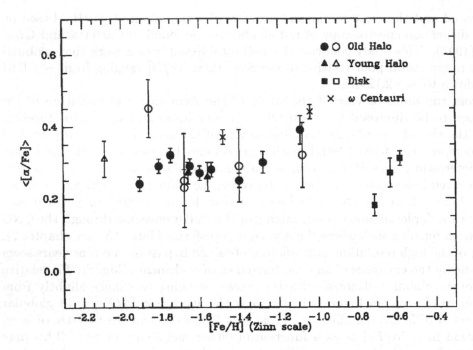

Figure 1.10. < [α/Fe] > vs. [Fe/H] for the globular clusters. Solid symbols represent clusters with three or more stars studied; open symbols represent clusters with only two stars analyzed. From Carney 1996, P.A.S.P. Vol. 108, 900; reproduced by kind permission of B. Carney and the University of Chicago Press (copy right 1996).

a) the color of the RGB at the horizontal branch $(B - V)_{0,g}$, which is the most widely used, and $(V - I)_{0,g}$. However, this index is strongly sensitive to the assumed reddening and uncertainties in the photometric calibrations. A calibration is then necessary to obtain the $[Fe/H]$ values; this is normally done by using $[Fe/H]$ ratios measured in spectroscopically bright field giants. b) The slope of the RGB, which is measured by $\Delta V_{1.4}$, namely the difference in V magnitude between the horizontal branch and the level of the RGB at the de-reddened color $(B - V)_0 = 1.4$

- *Integrated indices*: integrated photometry of clusters is a fast and sensitive way to derive information about their metal abundance, because it allows one to measure the abundances in very distant and faint clusters. However, owing to the fact that the whole cluster population contributes to the light, it is not easy to derive the weight of different chemical elements in the integrated light. Therefore, these photometric indices describe some average metallicity although they refer to features due to particular chemical elements such as Mg and Fe. The most common indices related to these two elements are the index M_{g_2} and the index $< Fe >$; each line index is a measurement of the flux contained in a wavelength region centered on the feature relative to the fluxes contained in red and blue continuum regions close to the feature (e.g. Faber et al. 1985). In particular, M_{g_2} refers to absorption lines due to molecules such as MgH and is defined as the difference, in magnitude, between the instrumental flux defined in a window centered on the line of Mg and the pseudo-continuum obtained by interpolation between two windows, one at the red side and the other at the blue side of the line. The Fe index instead, refers to atomic absorption lines and is defined in a similar way to the M_{g_2} but is expressed in terms of differences between wavelengths (namely in Angstrom). These indices are largely used to study the properties of the stellar populations in elliptical galaxies. The main problem with such indices is that they depend not only on the metallicity of the stars contributing to the integrated spectrum but also on their ages: this problem is known as *the age-metallicity* degeneracy. Once the indices are measured, one has to convert them into a measure of the metallicity, usually indicated by [Fe/H], and to do that one needs a calibration correlating the indices to the metallicity. This is done by adopting objects, such as globular clusters, where the metallicity can be measured in an independent way. The problem arises to calibrate the high metallicity range where independent measurements of metallicity are lacking and one should adopt theoretical models.

For galactic globular clusters a quite accurate abundance indicator is Q_{39} defined by Zinn (1980), which is substantially a measure of the strength of the Ca II H and K lines in the integrated cluster spectra.

- *RR-Lyrae indices.* The ΔS method is applied to the spectra of RR-Lyrae variables. It consists in measuring the strength of the Ca II lines using H lines to define temperature. The ΔS parameter can then be calibrated in terms of $[Fe/H]$ by using field RR Lyrae stars.

1.4   THE METALLICITY DISTRIBUTION OF THE LOCAL DISK G-DWARFS

The so-called "G-dwarf problem" (van den Bergh 1962; Schmidt 1963; Tinsley 1980) refers to the fact that the simple closed-box model of galactic chemical evolution predicts far too many metal poor disk stars than observed (see chapter 4). The G-dwarfs, namely the stars with lifetimes of the order of or larger than the age of the Galaxy, are important because they trace the star formation history of the local disk. Therefore, the G-dwarf metallicity distribution represents one of the most important observational constraints to models of galactic chemical evolution. The first accurate data on G-dwarfs were obtained by Pagel and Patchett (1975). Sommer-Larsen (1991) applied a correction to the distribution of Pagel and Patchett (1975) taking into account the fact that metal poor stars are older and have larger scale heights than metal-rich stars. Very recently, Wyse and Gilmore (1995) and Rocha-Pinto and Maciel (1996) have derived new metallicity distributions using *uvby* photometry and up-to-date parallaxes. These data again confirm the existence of a "G-dwarf problem" but differ considerably from the classic distribution of Pagel and Patchett. The different G-dwarf distributions are shown in Figure 1.11. In particular, the new data show a prominent peak around [Fe/H]=-0.2 dex which was not present in the previous distributions. Rana and Basu (1990) had also obtained a distribution with a marked peak, but the most recent ones do not confirm their finding of a large number of stars with [Fe/H]>0.1 dex. It is worth noting that none of the observed G-dwarfs belong to the galactic halo, since their positions and motions establish their disk membership.

1.5 THE METALLICITY DISTRIBUTION OF HALO STARS

The first metallicity distribution for the halo stellar population was obtained by Hartwick (1976) who studied the globular cluster system. He observed, in contrast to the disk G-dwarf metallicity distribution, that there is an excess of metal weak clusters compared with the simple model of galactic chemical evolution. He suggested that removal of gas from the star forming region (something like a galactic wind) could explain this excess of metal poor stars. In addition, the gas lost from the halo should have accumulated in the disk thus solving at the same time the G-dwarf problem, since the disk would have formed by already enriched gas. However, the current belief is rather more complicated and we return to this in the next chapters. More recent work on the metallicity distribution of halo stars is from Ryan and Norris (1991) who collected a large sample of kinematically selected halo stars with spectroscopically determined abundances. The metallicity distribution of halo stars is reported in Figure 1.12 and shows that there is a peak in metallicity at around [Fe/H] $\approx$ -1.6 dex. It is worth noting that this marked difference between the metallicity distribution of disk and halo stars indicates that different evolutionary histories took place in these different galactic components.

1.6 THE AGE-METALLICITY RELATION

As one might expect, the data generally show a decrease of the stellar metallicity with increasing galactic age, indicating a continuous growth of metals in the ISM during the life of the Galaxy (although the logarithmic nature of [Fe/H] produces the impression of a flattening of the metal content, see figures 1.13 and 1.14). Edvardsson et al. (1993) showed a plot of the iron abundance versus relative ages for the disk stars. The striking feature of this plot, shown in Figure 1.14, relative to previous ones, shown in Figure 1.13, is the large spread, especially for stars of intermediate ages. This spread is certainly physical since the mean

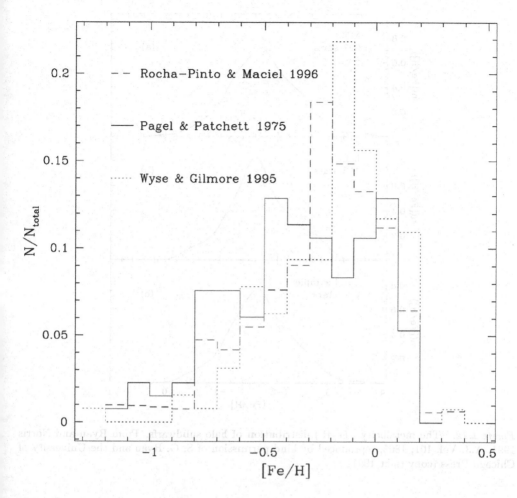

Figure 1.11. The G-dwarf metallicity distribution from different authors. Data sources are indicated in the figure. From Chiappini et al. 1997, Ap.J. Vol. 477, 765; reproduced by kind permission of C. Chiappini and the University of Chicago Press (copy right 1997).

errors in [Fe/H] and logarithmic age are about 0.1 dex. The origin of this scatter is not yet clear; orbital diffusion coupled with differential disk evolution (namely abundance gradients), inhomogeneous infall and local chemical inhomogeneities have been so far suggested. In particular, the fact that the observed spread could be due to orbital diffusion (Wielen 1977) represents a fascinating hypothesis. In this framework, the solar anomaly, namely the fact that the Sun has a metallicity larger by $+0.17 \pm 0.04$ dex than the average metallicity of nearby stars of solar age, can be explained by assuming that the Sun formed at a smaller galactocentric distance ($R_{i,\odot} = 6.6 \pm 0.9$ kpc) than the present one ($R_\odot \sim 8.5$ kpc), as

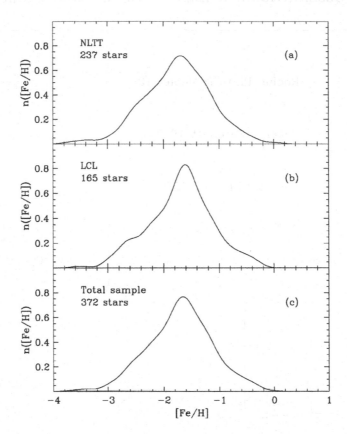

Figure 1.12. The metallicity ([Fe/H]) distribution of halo subdwarfs. From Ryan and Norris 1991, A.J. Vol. 101, 1865; reproduced by kind permission of S. G. Ryan and the University of Chicago Press (copy right 1991).

suggested by Wielen et al. (1996). On the other hand, the inhomogeneous infall might also explain the fact that the spread is larger in the absolute abundances than in the abundance ratios. Another possibility to explain at least part of the spread is that stars belonging to kinematically different galactic substructures such as halo, thick-disk, thin-disk and bulge can overlap in the solar neighbour-hood. Very likely, these different substructures have had different age-metallicity relations and this could explain part of the spread. Strobel (1991) tried to dis-entangle the age-metallicity relations for the thick-disk from that for the thin-disk. He collected data on both globular and open clusters and concluded that the stars of the thick-disk show a steeper rise of metallicity with time than the stars of the thin-disk, thus indicating two different histories of star formation; in particular, that the star formation proceeded faster in the thick than in the thin disk.

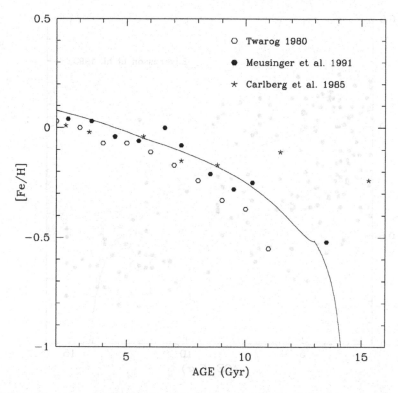

Figure 1.13. The age-metallicity ([Fe/H]) distribution for local disk stars. Data sources are indicated on the figure. The continuous line represents the prediction of the Chiappini et al. model. From Chiappini et al. 1997, Ap.J. Vol. 477, 765; reproduced by kind permission of C. Chiappini and the University of Chicago Press (copy right 1997).

1.7   ABUNDANCES IN THE BULGE STARS

The Milky Way bulge is a region within 1–2 kpc from the center, where a metal-rich rotationally supported stellar population resides. Rich (1988) determined [Fe/H] for 88 K giants in Baade's window and showed that the central stellar bulge of the Milky Way consists of a dominant stellar population with an age of ~10 Gyr and metallicity ranging from -1.5 to +1.0 dex with a peak located between +0.15 and +0.3 dex. This result was later confirmed by Geisler and Friel (1992) who used Washington photometry to derive the stellar metallicity. Concerning the abundance ratios in the galactic bulge, the first paper was by Barbuy and Grenon (1990) who found an average $< [O/Fe] > = +0.2$ dex for some super metal rich stars which they attributed to the bulge. McWilliam and Rich (1994) measured abundance ratios in metal-rich stars in the bulge, in the metallicity range $-1.08 < [Fe/H] < +0.44$. They found a somewhat puzzling pattern: Mg and Ti are enhanced by $\simeq 0.3$ dex relative to Fe over the whole

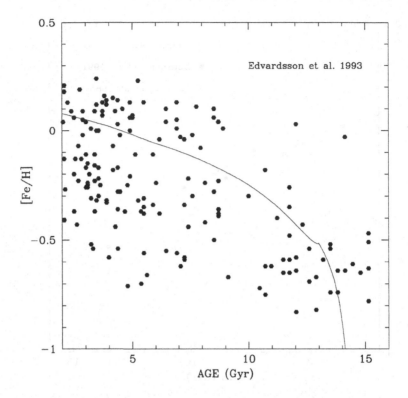

Figure 1.14. The age-metallicity ([Fe/H]) distribution for local disk stars. Data from Edvardsson et al. (1993). From Chiappini et al. 1997, Ap.J. Vol. 477, 765; reproduced by kind permission of C. Chiappini and the University of Chicago Press (copy right 1997).

[Fe/H] range. By contrast, Ca and Si closely follow the trend of disk stars, i. e. they reach solar ratios for [Fe/H]>-0.2 dex. They also found [Fe/H] values systematically lower by 0.25-0.3 dex than the values found by Rich (1988) (see Figure 1.15). One of the reasons for that could be that Rich derived his [Fe/H] values from the magnesium index M_{g_2} and assumed [Mg/Fe]=0.0 in his stars when passing from M_{g_2} to [Fe/H]. These lower values for [Fe/H] indicate that the galactic bulge is not so metal-rich as previously thought and that the average $< [Fe/H] >$ is roughly solar.

More recently, Sadler et al. (1996) found an average $< [Fe/H] > =-0.11 \pm 0.04$ dex for a sample of M and K giants in the bulge. They also found that stars with [Fe/H] <0 have [Mg/Fe] $\sim$ +0.2 dex whereas stars more metal rich do not show a Mg overabundance. Therefore, the situation still seems unclear although recent data by Barbuy (1999) seem to show overabundances for most of the α-elements observed in stars belonging to two bulge globular clusters. In Table 1.2 is shown a comparison between the abundance ratios in the two globular clusters NGC6553 and NGC6528 and bulge field stars. This table shows that both

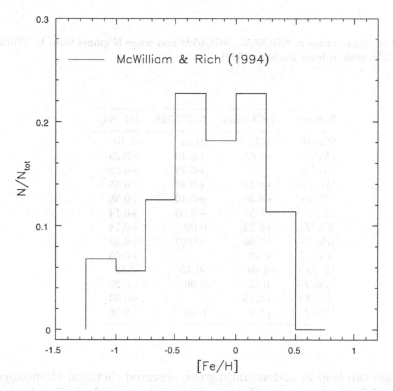

Figure 1.15. The metallicity distribution of bulge stars. Data from Mc William and Rich (1994).

globular clusters and field stars in the bulge indicate a general overabundance of α-elements relative to Fe, with individual variations from element to element and from globular cluster to field stars. For magnesium, in particular, the agreement between various objects is quite good.

Minniti et al. (1995) determined abundances for a large number of K giants at 1.5-1.7 kpc from the galactic center, based on medium resolution spectra. They concluded that there is a metallicity gradient within the inner 2 kpc of the Galaxy, although at any given distance from the galactic center there is a large spread in metal abundances. They also concluded that the presence of a metallicity gradient together with a correlation between metal abundance and kinematics of stars indicates the signature of dissipation in the process of formation of the galactic bulge (see next chapters), at variance with other suggestions implying that the bulge might have formed either from material ejected from the disk at late times or through dissipationless stellar merging. However, the existence of an abundance gradient in the bulge still needs to be confirmed.

Table 1.2. Abundance ratios in NGC6553, NGC6528 and bulge K giants from McWilliam and Rich (1994). The table is from Barbuy (1999).

Element	NGC6553	NGC6528	MR(94)
$[Fe/H]$	-0.55	-0.60	-0.40
$[Na/Fe]$	+0.65	+0.40	+0.20
$[O/Fe]$		+0.35	+0.03
$[Mg/Fe]$	+0.33	+0.40	+0.35
$[Al/Fe]$	+0.50	+0.40	+0.58
$[Si/Fe]$	+0.35	+0.60	+0.14
$[Ca/Fe]$	+0.32	0.00	+0.14
$[Ti/Fe]$	+0.60	+0.60	+0.37
$[Y/Fe]$	-0.20		+0.05
$[Zr/Fe]$	-0.40	-0.40	-0.53
$[Ba/Fe]$	-0.10	0.00	+0.20
$[La/Fe]$	+0.13		+0.09
$[Eu/Fe]$	+0.0	0.00	+0.26

Stellar ages can help in understanding the observed chemical abundances and the history of formation and evolution of the galactic bulge: Terndrup (1988) derived old ages ($\tau_{turn-off} \sim 11 \pm 3$ Gyr) from Color-Magnitude diagrams of M giants and concluded that bulge stars formed almost simultaneously and that there has been no star formation during the last 5 Gyr. Lee (1992) studied the bulge horizontal branch stars and concluded that the oldest stars in the bulge are even older than the oldest stars in the halo by 1.3 ± 0.3 Gyr. More recently, Ortolani et al. (1995) reached the conclusion that bulge globular clusters are at least as old as halo globular clusters. Therefore, there is a general consensus in attributing old ages to most of the bulge stars.

1.8   ABUNDANCES FROM PLANETARY NEBULAE

Element abundances for Planetary Nebulae (PNe) are based on optical data and therefore are restricted to regions within few kpc of the Sun. The spectra of PNe show strong forbidden lines (especially [OIII]4949, 5007 Å) on a weak continuum. The abundances of elements manufactured inside the central star (He, C, N) give us information about the nucleosynthesis processes and enable us to put constraints on nucleosynthetic theories, whereas the abundances of elements that are not manufactured inside the star (O, S, Ne) can be used to trace the galactic chemical evolution. There are five basic types of PNe, as defined originally by Peimbert (1978) with some later refinements, associated with different stellar populations in the Galaxy. This classification is related to the evolution of intermediate mass stars ($0.8 \leq M/M_\odot \leq 8$), which are the progenitors of PNe. The sequence of types I, II, III, and IV corresponds approximately to a sequence of decreasing progenitor masses. From the chemical point of view, type I PNe show N and He excesses, type II show little or no excess in these ratios and they are closely associated with the galactic disk (they can be further subdivided in type IIa and IIb according to whether they are N-rich or contain very little N, respectively); type III PNe have strong metal underabundances and large peculiar velocities; type IV are halo objects and they also show strong metal deficiencies. Finally type V PNe are bulge objects. The determination of the abundances of those elements which are not affected by stellar evolution and nucleosynthesis in type II PNe should reflect the abundances of the ISM along the galactic disk and be similar to the abundances derived from HII regions. The derived galactic disk gradients have the form $log(X/H) + 12 = aR + b$, where a and b are constants. They are derived for a galactocentric distance range between 5 and 12 kpc, with the Sun at 8.5 kpc from the galactic center. The gradient of O derived from PNe gives $a \simeq -0.069$ dex kpc^{-1}, a very similar value to that derived from the HII regions $a \simeq -0.07$ dex kpc^{-1} (see later). The oxygen data from PNe, HII regions and B type stars are compared in figure 1.17.

The value of a for Ne in PNe is slightly lower than that for O and the value of a for S is very similar to that of O. A further interesting point is the comparison between the gradients derived from different types of PNe. In particular, the type III PNe, related to the oldest stars, show a flatter gradient than the type II, I PNe, related to the younger objects. This can be interpreted in two ways, either the oldest stars have moved away from their original birthplaces thus washing out the original gradient or the abundance gradient evolved with time gradually becoming steeper. However, the opposite effect seems possibly evident through gradients derived from old and young open clusters (Carraro et al. 1998a). Therefore, the question of whether the abundance gradients have steepened or flattened with time is still open, also because in both cases (PNe and open clusters) dynamical effects could have played a role in the actual radial distribution of these objects.

1.9   ABUNDANCES FROM HII REGIONS, B STARS, OPEN CLUSTERS

Recent reviews on abundances in galactic HII regions are by Wilson and Rood (1994), Peimbert (1993) and Wilson and Matteucci (1992). There are two types of abundance determinations relative to hydrogen in HII regions: one is based on recombination lines which should have a weak temperature dependence of the nebulae (He, C, N, O) and the other is based on collisionally excited lines where a strong dependence is intrinsic to the method (C, N, O, Ne, Si, S, Cl, Ar, Fe and Ni); the second method has predominated until now. It is worth mentioning also in connection with abundance determinations in PNe that recent abundance determinations of C, N, O, Ne in a sample of PNe using faint recombination lines of these elements reveal large differences from the results using collisionally excited lines. The reasons for these discrepancies are not yet understood, but spatial variations of abundances within individual PNe seem to be well established (Liu et al. 2000).

A direct determination of the abundance gradients from HII regions in our Galaxy from the optical lines is difficult because of extinction. For HII regions more than 3 kpc distant, dust extinction is usually large and studies must be made in the infrared or radio wavelength regions. One of the problems in deriving abundances from HII regions is the fact the abundances should be derived relative to H and therefore one needs ionization models for the HII region. Shaver et al. (1983) derived abundance gradients from galactic HII regions by using optical as well as radio recombination lines. Their results showed pronounced gradients in N (-0.09 ± 0.015) and O (-0.07 ± 0.015) and a very flat sulphur gradient (see also Table 1.3). These results have been extended by Fich and Silkey (1991) who determined abundances in outer parts of the northern Milky Way and showed that the N gradient is flatter than that derived by the previous authors. However, Simpson et al. (1995) who derived galactic abundance gradients from far infrared lines found gradients comparable to those of Shaver et al. except for sulphur which is found to have a gradient as steep as that of oxygen. In particular, they found that N has the steepest gradient (a=-0.10 ± 0.02 dex kpc^{-1}), followed successively by oxygen, then neon and then sulphur. However, they suggested that their data can also be fitted by a pattern different from a unique monotonic gradient: in fact, they can be fitted by a pattern in which the Galaxy is divided into 2 or 3 regions with abundances decreasing in steps from the inner region to the outer. As has been noted in previous studies, they found that the N/O ratio derived from far infrared lines is roughly a factor of two higher than that derived from optical lines, but the cause of this difference is not yet well understood. Vilchez and Esteban (1996) performed optical and near infrared spectroscopy for a sample of HII regions in the galactic anticenter. They derived the abundances of He, N, O and S and found that the gradients are nearly flat out to 18 kpc, suggesting the existence of two different slopes for the inner (steep) and outer (flat) gradients in the galactic disk, where the inner region is defined for galactocentric distances $R_G \leq 6$ kpc. However, this behaviour has not been confirmed by other studies, and a unique slope for the gradients cannot be ruled out. Figure 1.16 shows the abundance gradients derived from HII regions.

Gradients of different chemical elements are in principle expected to be different from one another as a result of the nucleosynthesis and stellar progenitors. For example, the existence of a gradient in N/O indicates the different nature

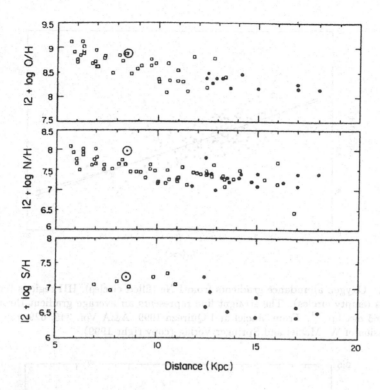

Figure 1.16. Radial abundance gradients from HII regions. Data from Shaver et al. (1983), Fich and Silkey (1991) and Vilchez and Esteban (1996). From Pagel 1997, "Nucleosynthesis and Chemical Evolution of Galaxies"; reproduced by kind permission of B.E.J. Pagel and the Cambridge University Press (copy right 1997).

and nucleosynthesis processes undergone by these two elements and discussed in succeeding chapters.

Abundance gradients along the galactic disk can be inferred also from abundance measurements in B stars and open clusters. Kaufer et al. (1994) found that B type stars did not exhibit any appreciable gradient but more recent studies (Smartt and Rolleston, 1997 and Gummersbach et al. 1998) found a gradient of oxygen of -0.07 ± 0.01 dex kpc^{-1} between 6 and 18 kpc and -0.07 ± 0.02 dex kpc^{-1} between 5 and 14 kpc, respectively. These last results agree with the gradient estimated from HII regions and with the planetary nebula data (see figure 1.17).

Open clusters seemed to indicate (Friel 1995) that the iron abundance [Fe/H] shows a steep gradient of -0.095 ± 0.017 dex kpc^{-1} between 7 and 16 kpc but Friel (1999) in a more recent determination found a [Fe/H] gradient shallower than the previous one and closer to the O gradients estimated recently from B

36

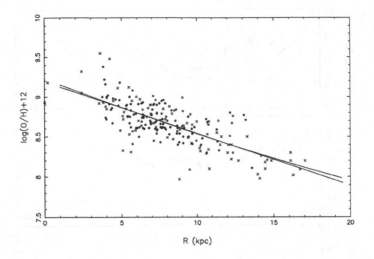

Figure 1.17. Oxygen abundance gradients from PNe (filled circles), HII regions (crosses) and B type stars (empty circles). The straight line represents an average gradient through all the data of -0.065 dex kpc^{-1}. From Maciel and Quireza 1999, A&A Vol. 345, 629; reproduced by kind permission of W. Maciel and Springer Verlag (copy right 1999).

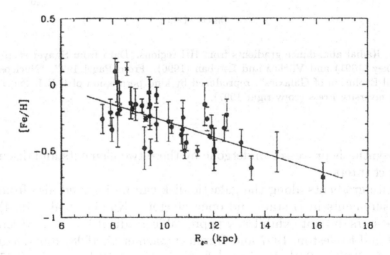

Figure 1.18. The radial gradient of [Fe/H] as measured from open clusters. Figure from Friel 1999, Ap. & S.S. Vol. 265, 271; reproduced by kind permission of Kluwer Academic Publishers (copy right 1999).

type stars (see Figure 1.18). In Table 1.3 we report a large compilation of data on abundance gradients in the galactic disk by both old and recent studies.

Table 1.3. Summary of abundance gradients along the galactic disk.

Element	Study	Gradient (dex kpc^{-1})	$R_\odot$ (kpc)	Radial Baseline (kpc)	Object
Oxygen	Shaver et al. 1983	−0.07 ± 0.015	10	5—13	H II regions
	Gehren et al. 1985	−0.01 ± 0.02		8—18	B-type stars
	Fitzsimmons et al. 1992	−0.03 ± 0.02	8.5	6—13	B-type stars
	Kaufer et al. 1994	−0.000 ± 0.009	8.5	6—17	B-type stars
	Kilian-Montenbruck et al. 1994	−0.021 ± 0.012	8.7	6—15	B-type stars
	Maciel & Köppen 1994	−0.069 ± 0.006	8.5	4—13	PNe
	Vílchez & Esteban 1996	−0.036 ± 0.020	8.5	12—18	H II regions
	Afflerbach et al. 1997	−0.064 ± 0.009	8.5	0—12	H II regions
	Smartt & Rolleston 1997	−0.07 ± 0.01	8.5	6—18	B-type stars
	Gummersbach et al. 1998	−0.067 ± 0.024	8.5	5—14	B-type stars
	Maciel & Quireza 1999	−0.058 ± 0.007	7.6	3—14	PNe
Nitrogen	Shaver et al. 1983	−0.09 ± 0.015	10	5—13	H II regions
	Kaufer et al. 1994	−0.026 ± 0.009	8.5	6—17	B-type stars
	Kilian-Montenbruck et al. 1994	−0.017 ± 0.020	8.7	6—15	B-type stars
	Simpson et al. 1995	−0.10 ± 0.02	8.5	0—10	H II regions
	Vílchez & Esteban 1996	−0.009 ± 0.020	8.5	12—18	H II regions
	Afflerbach et al. 1997	−0.072 ± 0.006	8.5	0—12	H II regions
	Rudolph et al. 1997	−0.111 ± 0.012	8.5	0—17	H II regions
	Gummersbach et al. 1998	−0.078 ± 0.023	8.5	5—14	B-type stars
Sulphur	Shaver et al. 1983	−0.01 ± 0.020	10	5—13	H II regions
	Kilian-Montenbruck et al. 1994	−0.026 ± 0.025	8.7	6—15	B-type stars
	Maciel & Köppen 1994	−0.067 ± 0.006	8.5	4—13	PNe
	Simpson et al. 1995	−0.07 ± 0.02	8.5	0—10	H II regions
	Vílchez & Esteban 1996	−0.041 ± 0.020	8.5	12—18	H II regions
	Afflerbach et al. 1997	−0.063 ± 0.006	8.5	0—12	H II regions
	Rudolph et al. 1997	−0.079 ± 0.009	8.5	0—17	H II regions
	Maciel & Quireza 1999	−0.077 ± 0.011	7.6	3—14	PNe
Neon	Kilian-Montenbruck et al. 1994	−0.043 ± 0.011	8.7	6—15	B-type stars
	Maciel & Köppen 1994	−0.056 ± 0.007	8.5	4—13	PNe
	Simpson et al. 1995	−0.08 ± 0.02	8.5	0—10	H II regions
	Maciel & Quireza 1999	−0.036 ± 0.010	7.6	3—14	PNe
Iron	Friel & Janes 1993	−0.09 ± 0.02	8.5	7—16	Open Clusters
	Kilian-Montenbruck et al. 1994	−0.003 ± 0.020	8.7	6—15	B-type stars
	Twarog et al. 1997	−0.067 ± 0.008	8.5	6—16	Open Clusters
	Friel 1999	−0.06 ± 0.01	8.5	7—16	Open Clusters
	Carraro et al. 1998a	−0.09	8.5	7.5—16	Open clusters
Helium	Shaver et al. 1983	−0.001 ± 0.008	10	5—13	H II regions
Carbon	Kilian-Montenbruck et al. 1994	+0.001 ± 0.015	8.7	6—15	B-type stars
	Gummersbach et al. 1998	−0.035 ± 0.014	8.5	5—14	B-type stars
Magnesium	Kilian-Montenbruck et al. 1994	−0.020 ± 0.011	8.7	6—15	B-type stars
	Gummersbach et al. 1998	−0.082 ± 0.026	8.5	5—14	B-type stars
Silicon	Kilian-Montenbruck et al. 1994	+0.000 ± 0.018	8.7	6—15	B-type stars
	Gummersbach et al. 1998	−0.107 ± 0.028	8.5	5—14	B-type stars

1.10 ISOTOPIC RATIOS

Isotopic abundance ratios are derived from radio and mm observations of molecules CO, CH_2O, HCN, NH_3. The $^{13}C/^{12}C$, $^{14}N/^{15}N$, $^{16}O/^{18}O$ and $^{18}O/^{17}O$ ratios in the solar system, local ISM and galactic center are reported in Table 1.4. The observed abundance gradients as functions of the galactocentric distance for the same isotopes are shown in Figure 1.19. Such gradients of isotopic ratios are very useful to infer the nucleosynthetic origin of the elements, such as their *primary* or *secondary* origin, to be discussed later.

Table 1.4. Isotopic ratios in the Galaxy. Data taken from Wilson and Rood (1994)

Isotope	Galactic center	Local interstellar medium	Solar System
$(^{12}C/^{13}C)$	~20	76 ± 7	89
$(^{14}N/^{15}N)$	> 600	370 ± 23	270
$(^{16}O/^{18}O)$	250	560 ± 25	490
$(^{18}O/^{17}O)$	3.2 ± 0.2	3.2 ± 0.2	5.5

1.11 THE ABUNDANCES OF LIGHT ELEMENTS

Table 1.5. Abundances by number of light elements

Element	Primordial abundance	ISM abundance	Solar System abundance	Source
D/H	$(3.4 \pm 0.3) \cdot 10^{-5}$ $-(2.0 \pm 0.5) \cdot 10^{-4}$	$(1.6 \pm 0.1) \cdot 10^{-5}$	$(2.6 \pm 0.6) \cdot 10^{-5}$	1,2, 3,4
$^3He/H$	$(1-2) \cdot 10^{-5}$	$(2 \pm 1) \cdot 10^{-5}$	$1.80 \cdot 10^{-5}$	5,6,11
$^4He/H$	0.0760-0.0795	0.1071	0.100	6,7,12,13
$^7Li/H$	$(1.73 \pm 0.21) \cdot 10^{-10}$	$3.16 - 6.76 \cdot 10^{-9}$	$1.97 \cdot 10^{-9}$	8,9,6
$^6Li/H$	$(2-9) \cdot 10^{-14}$		$1.59 \cdot 10^{-10}$	10,6
$^9Be/H$	$\sim (0.04-2) \cdot 10^{-17}$		$2.73 \cdot 10^{-11}$	6,10
$^{10}B/H$	$\sim (0.5-3) \cdot 10^{-19}$		$1.49 \cdot 10^{-10}$	6,10
$^{11}B/H$	$\sim (0.02-1) \cdot 10^{-16}$		$6.39 \cdot 10^{-10}$	6,10

1. Burles and Tytler (1998); 2. Webb et al. (1997); 3. Linsky et al. (1993;1995); Scully et al. (1996); 5. Gloeckler and Geiss (1996); 6. Grevesse and Noels (1993); 7. Isotov et al. (1999); 8. Bonifacio and Molaro (1997); 9. Duncan and Rebull (1996); 10. Thomas et al. (1993); 11. Olive and Fields (1999); 12. Olive et al. (1997); 13. Chiappini et al. (1997).

The light elements D, ^3He, ^4He and ^7Li are thought to be produced during the Big Bang and also during the galactic evolution, with the exception of D which is only destroyed inside stars. The primordial abundances of these elements are

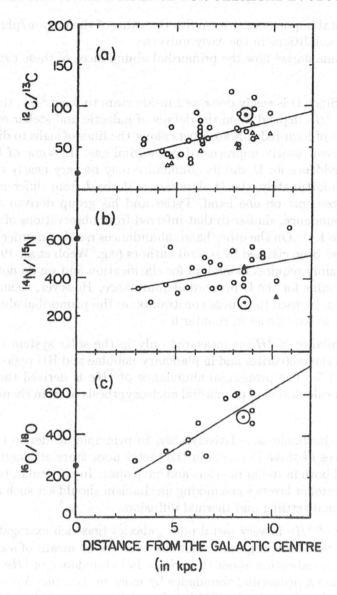

Figure 1.19. Variation of CNO isotopic ratios with galactocentric distance deduced from mm wave measurements of molecules in molecular clouds. (a) $^{12}C/^{13}C$ from CO(triangles) and formaldehyde CH_2O (circles). (b) $^{14}N/^{15}N$ from HCN (circles) and NH_3 (triangles). (c) $^{16}O/^{18}O$ from formaldehyde. Solar-System values are indicated by the $\odot$ sign and galactic centre values (or a lower limit in the case of $^{14}N/^{15}N$) by a filled circle. Data from Wilson and Rood (1994). From Pagel 1997, "Nucleosynthesis and Chemical Evolution of Galaxies"; reproduced by kind permission of B.E.J. Pagel and the Cambridge University Press (copy right 1997).

of fundamental importance to establish the value of the baryon/photon ratio and the physical conditions in the early universe.

Here we summarize how the primordial abundances of these elements are derived:

- D, ^3He–Since D is easily destroyed inside stars to form ^3He, the abundances of D and ^3He depend upon the details of galactic and stellar evolution. Observations of high-redshift absorbers along the lines of sight to distant quasars (DLA) reveal nearly unprocessed primordial gas. In some of these systems there is evidence for D and its abundance may be very nearly the primordial value. Unfortunately, the D abundances derived from different authors are in disagreement: on one hand, Tytler and his group derived a low primordial D abundance, similar to that inferred from observations of the local ISM (see Table 1.4). On the other hand, abundances nearly an order of magnitude larger have been claimed by several authors (e.g. Webb et al. 1997). This dispute certainly requires more data for clarification and we do not suggest here any best value for the D primordial abundance. However, chemical evolution models can be used to impose constraints on the primordial abundances of D and ^3He, as we discuss in chapter 6.

  The abundance of ^3He is measured only in the solar system from the solar wind and the meteorites and in planetary nebulae and HII regions as reported in Table 1.5. The primordial abundance of ^3He is derived theoretically, either from calculation of primordial nucleosynthesis or from chemical evolution models.

- ^4He, ^7Li–It should be relatively easy, in principle, to derive the primordial abundances of these elements in the metal poor stars although ^7Li is easily destroyed both in stellar interiors and envelopes. In particular, to destroy Li in the outer stellar layers some mixing mechanism should act such as convection, gravitational settling and thermal diffusion.

  Measures of ^4He in very metal poor galaxies (gas-rich extragalactic HII- regions) or the study of galactic globular clusters by means of stellar evolution can give an indication about the primordial abundance of ^4He. These methods indicate a primordial abundance by mass in the range $Y_P = 0.23 - 0.248$. The ^4He abundance in the ISM is inferred from measurements in HII regions and from chemical evolution models.

  The primordial abundance of ^7Li is inferred from abundance analyses of Pop II stars which show no Li-depletion and a roughly constant value of the Li abundance. The Li abundance in disk stars shows a general increase but with a very big spread which is interpreted to be due to its destruction in stars according to their evolutionary phase and metallicity, whereby only the stars lying in the upper envelope of these data may trace the Li abundance of the

ISM at the time of their birth (see Figure 1.20). However, another view claims that the Li abundance observed in Pop II stars is the result of Li depletion, and that the true primordial abundance is that observed in the youngest stars (e.g. T-Tauri stars). In this case, a non-standard Big Bang nucleosynthesis is required to explain the high primordial Li abundance. The low primordial ^7Li abundance hypothesis is also supported by observations indicating that Li is produced in stars. In particular, the asymptotic-giant branch stars (AGB stars) are observed to be Li-rich:

- massive AGB (5-8 $M_\odot$) and M supergiants in the Magellanic Clouds. These stars present enhanced s-process elements from which one infers that they are in their thermal-pulsing phase, but they are not C-stars. Smith and Lambert (1989, 1990) found A(Li)$\simeq$ 2.0 − 3.8 dex (where A(Li)=logN(Li)/H +12) and Plez et al. (1993) A(Li)$\simeq$ 3.0 dex for these stars.

- galactic C-stars with luminosities from $M_{bol} \simeq$ -6 to -3.5 show in some cases very high Li abundances (A(Li) $\simeq$ 4.5-5.5 dex, Abia et al. 1993)

- ^6Li, Be, $B-$ These elements do not have a stellar origin and are formed as a result of α-α fusion and spallation reactions between protons, α-particles and (mainly) CNO nuclei at high energies. Their primordial abundances can, in principle, be measured in Pop II stars although the measure is quite difficult given the very small concentrations of these elements and they also probably contain already some contribution from spallation processes. Therefore, the values reported in Table 1.5 for the primordial abundances of these elements are those theoretically derived. A fraction of ^7Li is also likely to be produced by spallation although there is still a large uncertainty on the possible amount of ^7Li produced in this way. There are indications both theoretical and observational that cosmic-ray α-α reactions can at most contribute to 10 − 20% of the Pop I lithium (Walker et al. 1985; Prantzos et al. 1993; Reeves 1993). The simplest argument derives from the fact that the meteoritic ratio $(^7Li/^6Li)_{meteoritic} = 13$ is larger than the same ratio predicted by spallation processes $(^7Li/^6Li)_{CR} = 2.0$, thus indicating that most of the ^7Li in the solar system should have a stellar origin.

1.12   THE MASS DISTRIBUTION IN THE GALAXY
1.12.1   THE DISTRIBUTION OF GAS

In order to have an estimate of the total surface gas density along the galactic disk one should add the contributions from the neutral hydrogen (HI), the molecular hydrogen (H_2) and the associated helium.

The distribution of the HI surface mass density is reasonably well determined from 21 cm line observations, the main uncertainty residing in the optical depth

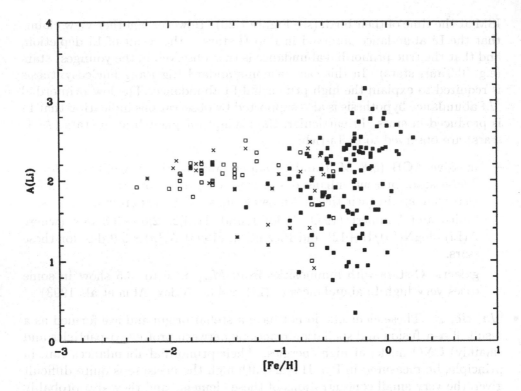

Figure 1.20. A(Li) vs. [Fe/H] as traced by stars in the solar neighbourhood, $A(Li) = \log (N(Li)/H) + 12$, where $N(Li)$ is the abundance by number. Filled squares: disk stars; empty squares: non-disk stars; asterisks and crosses: kinematically unidentified stars. Data and figure from Romano et al. 1999, A&A Vol. 352, 117; reproduced by kind permission of Springer Verlag (copy right 1999).

of the 21cm emission. It appears roughly constant with galactocentric distance $\sim (4-5)M_{\odot}pc^{-2}$, in the range $4 \leq R_G \leq 8.5$ kpc. For the HI outside the solar circle the biggest uncertainty results from the adoption of different rotation curves (flat or slowly rising). On the other hand, the molecular component is poorly known and is inferred from measurements of z-integrated CO emissivity, I_{CO}, using a conversion factor $C = N_{H_2}/I_{CO}$ which is often assumed to be independent of the galactocentric distance. This may not be correct since there is evidence for abundance gradients along the disk and this is perhaps the major cause of uncertainty. Generally, three different methods are used to derive C: the first method is based on measuring the extinction of stars behind nearby molecular clouds, the second is based on gamma-ray counts to estimate hydrogen column densities and the third estimates the masses of the molecular clouds from the virial theorem. It is worth noting that the uncertainty in the CO emissivity alone can be as much as a factor of two. The distribution of the H_2 derived in this way shows that the molecular hydrogen follows the starlight distribution, namely is falling exponentially with increasing galactocentric distance, with a pronounced molecular ring at 5 kpc from the galactic center. After taking into account all of the gaseous components, one finds that the total surface gas density $(HI + H_2)$ of the disk follows an exponential law with a scale length of 5 ± 1 kpc over the range $5 \leq R_G \leq 20$ kpc but with a much lower value in the range $1 \leq R_G \leq 5$ kpc. In Figure 1.21 we show a compilation of observational determinations of the total gas surface mass density $(HI + H_2)$ along the galactic disk, whereas in Figure 1.22 the distributions of HI and H_2 are shown separately.

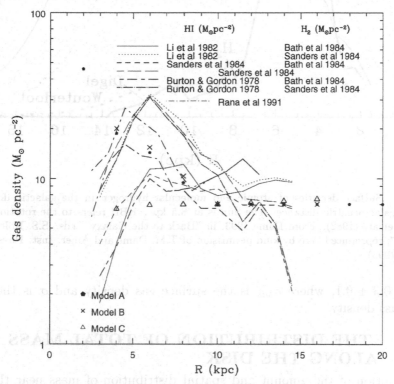

Figure 1.21. The distribution of the total surface gas density along the galactic disk: each line represents the sum of HI and H_2 from the sources indicated in the figure. The Sun is at 8.5 kpc. The model predictions and the figure are from Matteucci and Chiappini 1999, in "Chemical Evolution from Zero to High Redshift", eds. J.R. Walsh and M.R. Rosa, p.83; reproduced by kind permission of Springer Verlag (copy right 1999).

The surface mass density of gas in the solar neighbourhood is $\sigma_{ISM} = 13 \pm 3 M_\odot pc^{-2}$ comprising $10 M_\odot pc^{-2}$ of atomic gas and $3 \pm 3 M_\odot pc^{-2}$ of molecular hydrogen. Considering that the local total surface mass density of the disk, estimated from the vertical stellar motions reflecting the galactic potential, is $\simeq 50 M_\odot pc^{-2}$ (see next section), one can derive a local gas fraction of $\mu =$

44

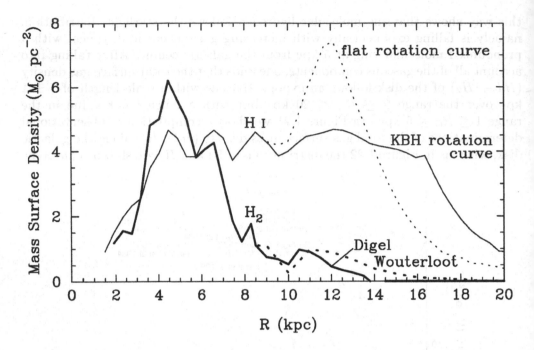

Figure 1.22. Surface densities of atomic and molecular hydrogen in the galactic disk as a function of galactocentric distance. The Sun is at 8.5 kpc. KBH refers to the rotation curve of Kulkarni et al. (1982). From Dame 1993, in "Back to the Galaxy", eds. S.S. Holt and F. Verter, p.267; reproduced here by kind permission of T.M. Dame and Amer. Inst. Phys. Publ. (copy right 1993).

$\sigma_{gas}/\sigma = 0.3 \pm 0.1$, where σ_{gas} is the surface gas density and σ is the total surface mass density.

1.12.2 THE DISTRIBUTION OF TOTAL MASS ALONG THE DISK

A description of the amount and spatial distribution of mass near the Sun is an essential ingredient for studying the dynamical and chemical evolution of the Galaxy. For the total mass discussed here and in the following we utilize the dynamical mass which contains all the baryonic matter plus eventually some unaccounted matter (M_{dark}):

$$M_{tot} = M_{stars} + M_{remnants} + M_{gas} + M_{dark} \tag{1.2}$$

where M_{stars} indicates the mass of living stars (including brown dwarfs), $M_{remnants}$ indicates the mass in the form of white dwarfs, neutron stars and black holes and M_{gas} is the total mass of the interstellar gas. The distribution of mass in the galactic disk is characterized by two numbers, the local volume density ρ_o

(known as "Oort limit") and the local total surface mass density σ_o. The former has units of $M_\odot pc^{-3}$ whereas the latter has units of $M_\odot pc^{-2}$. The best available determination of local volume density of *identified* material is $\rho_{obs} \sim 0.1 M_\odot pc^{-3}$ with an uncertainty of $\simeq 25\%$. Comparison of this value with that determined from dynamical analyses is required to test the existence of dark matter (missing mass) in the galactic disk. In order to measure the amount of mass one needs to know the galactic gravitational vertical force, $K_z(z)$, which is done by studying the vertical motion of stellar samples. This gives the vertical component of the gravitational force which combined with the already known component in the direction of the circular motion allows us to derive the galactic potential and through this potential the volume gas density by applying the Poisson equation (see more specialized tests for a detailed treatment). Both F dwarf and K giant tracer samples have been analyzed to derive ρ_o, with the result that $\rho_o \sim 0.20 M_\odot pc^{-3}$ (Bahcall 1984), thus suggesting that there should be at least as much dark matter as luminous matter in the solar neighbourhood, a result already found by Oort. The main uncertainties in such a derivation of ρ_o rest on the modelling of the stellar velocity distribution near the galactic plane, and the determination of the stellar density distribution with distance from this plane. Kuijken and Gilmore (1989 I, II, III, 1991) re-examined the derivation of the Oort limit and showed that the analysis of Bahcall was dependent on the stellar sample adopted to derive the galactic potential, and that the available F star and K giant data are not able to provide any evidence in favor or against the existence of a missing mass in the galactic disk. However, the important point of their work was that they were able to derive the total galactic potential below an effective distance from the plane of 1.1 kpc, and could relate this potential to the integral surface mass density instead than to the volume mass density. They found that the value of the potential is equivalent to $71 \pm 6 M_\odot$ pc^{-2} which is the surface mass density corresponding to the total amount of mass in a column perpendicular to the galactic plane. This is what is required to interpret the rotation curves and the large scale distribution of mass in galaxies. By adopting the galactic rotation curve, Kuijken and Gilmore were then able to separate the total surface mass density into two components (halo and disk), and obtained for the disk a value of $\sigma_o \sim 48 \pm 9 M_\odot$ pc^{-2}. This kind of analysis used a stellar sample of K dwarfs which matches the conditions required to determine the local surface mass density. In fact, the tracer stars should be in dynamical equilibrium with the galactic potential (i.e. stars sufficiently old to have completed many vertical oscillations through the galactic plane), should have reliable photometric and spectroscopic data, and should extend beyond most of the disk mass and have reliable distances.

They also obtained the observed total surface mass density in the solar neighbourhood by summing the surface mass density of stars, remnants (mainly white dwarfs) and interstellar gas, $\sigma_{obs} = 48 \pm 8 M_\odot pc^{-2}$. Therefore, the important conclusion of their work was that there is no real difference between the dynamical mass of the local disk and the observed mass (stars plus gas), and consequently no robust evidence for the existence of any unaccounted dark mass associated with the local galactic disk.

In models for the chemical evolution of the Galaxy one needs to know the distribution along the disk of the total surface mass density. Generally the mass distribution along the galactic disk can be approximated by an exponential distribution with a scale length of ~ 3.7 kpc (Bahcall, 1984):

$$\sigma(R) = \sigma_c e^{-R/R_d} \qquad (1.3)$$

where σ_c is the central total surface mass density and R_d is the scale length. It should be said that this scale length depends on the assumed stellar sample. For example, a larger scale length arises from the studies of old stellar populations, since they are found at a larger height from the galactic plane than young stars. The range in which the disk scale length can be defined is $R_d = 2.3 - 5.0$ kpc.

In Table 1.6 we report the radial total surface mass density distribution adopted by Rana (1991) who assumed the distribution of Bahcall but normalized it to $54 M_\odot pc^{-2}$ at the solar circle, in order to take into account the result of Kuijken and Gilmore.

Table 1.6.   Dynamical surface mass density along the galactic disk. Rana (1991).

Galactocentric distance R_G (kpc)	Dynamical mass σ ($M_\odot pc^{-2}$)
2.5	271.1
3.5	207.2
4.5	158.3
5.5	121.0
6.5	92.50
7.5	70.70
8.5	54.00
9.5	41.30
10.5	31.50
11.5	24.10
12.5	18.40

1.12.3   THE GALACTIC DARK MATTER HALO

We have seen in the previous section that the amount of unaccounted dark matter in the local galactic disk seems to be negligible. However, studies of the rotation curves in spirals, which are measured optically from emission lines in

HII regions or by using the 21 cm emission line of neutral hydrogen, show the existence of extended large halos of unseen matter. In particular, the rotation curves do not indicate any Keplerian fall-off of the rotational velocity at large galactocentric distances ($V(R_G) \propto R_G^{-1/2}$), as expected if the matter would follow the distribution of light, but rather a flattening ($V(R_G)$=constant). This is interpreted as due to an additional and invisible large mass component extending outside the distribution of visible stars towards very large galactocentric distances. This characteristic seems to be common also to our Galaxy although the galactic rotation curve can give us information on the galactic mass only up to $2R_\odot$ (where $R_\odot$ is the solar galactocentric distance). This is due to the lack of rotational velocity tracers at larger distances. Therefore the galactic rotation curve can give us an indication of the mass enclosed inside 20 kpc. In particular, when the rotation curve of the Galaxy is interpreted in terms of a mass model it shows that the total mass inside the central 20 kpc is $M_{tot} = 2 \cdot 10^{11} M_\odot$ (see Figure 1.23), which does not imply a large fraction of dark matter.

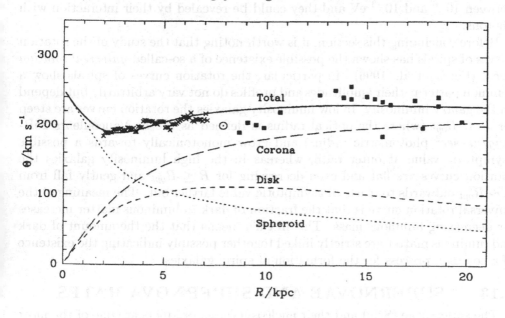

Figure 1.23. A three-component model of the Milky Way fitted to the rotation curve of the Galaxy. The curve labelled "corona" refers to the dark matter halo.From Merrifield 1992, A.J. Vol. 103, 1552; reproduced here by kind permission of M. Merrifield and the University of Chicago Press (copy right 1992).

To constrain the extent and the mass of the galactic dark halo we should adopt other methods such as the kinematics of globular clusters and local satellite galaxies. All the dynamical mass tracers suggest a minimal mass for the Galaxy and a maximum extension for the halo; in particular, Fich and Tremaine (1991) suggest $M_{tot} \sim 4 \cdot 10^{11} M_{\odot}$ and $r_{max} \sim 35$ kpc . More recent estimates of the mass of the Galaxy by means of kinematical data of satellites and globular clusters have been able to improve this estimate suggesting a total mass of $M_{tot} \sim 2 \cdot 10^{12} M_{\odot}$ extending far beyond a radius of ~ 200 kpc (Wilkinson and Evans, 1999). This mass is at least ten times larger than the visible mass in the disk and implies that the Milky Way does indeed possess an extended and massive dark halo. The nature of dark matter is still unknown although different suggestions have been made: planets, brown dwarfs, stellar remnants and molecular clouds for the baryonic dark matter and hypothetical objects such as neutralinos and axions for the non-baryonic dark matter. Neutralinos arise in theories of elementary particles involving supersymmetry (e.g. Sadoulet, 1999): they should weigh like a large atom, interact with normal matter only through weak interactions and be produced in the early universe. The axions were hypothesized in order to solve a problem in quantum chromodynamics and they can also be produced in the early universe (e.g. Sikivie, 1983); they are very light objects with a mass between 10^{-6} and 10^{-3} eV and they could be revealed by their interaction with photons.

Before concluding this section, it is worth noting that the study of the rotation curves of spirals has shown the possible existence of a so-called *universal rotation curve* (Persic et al. 1996). In particular, the rotation curves of spirals show a common pattern: their amplitudes and profiles do not vary arbitrarily but depend on the galaxy luminosity. In low luminosity galaxies the rotation curves are steep for $R \leq R_{opt}$ (where the optical radius is defined as the de Vaucouleurs 25B-$mag/arcsec^2$ photometric radius) and grow monotonically towards a possible asymptotic value at outer radii, whereas in the high luminosity galaxies the rotation curves are flat and even decreasing for $R \leq R_{opt}$ and gently fall from $R \sim R_{opt}$ outwards to reach an asymptotic value farther out. The meaning of the universal rotation curve is that the fraction of dark to luminous matter increases for decreasing luminous mass. This in turn means that the the amount of dark and luminous matter are strictly linked together possibly indicating the existence of a common process for the formation of spiral galaxies.

1.13 SUPERNOVAE AND SUPERNOVA RATES

The supernovae (SNe) and their nucleosynthesis products are one of the most important ingredients in models of chemical evolution . Here we will summarize the most important observational facts concerning supernovae.

1.13.1 SUPERNOVA CLASSIFICATION

According to the original Zwicky classification, supernovae are called type II if they show strong hydrogen emission lines in their spectra and type I otherwise.

Type I supernovae (SN I) are further subclassified into Ia, Ib and Ic. As shown in Figure 1.24 such subtypes are defined by the early time (~ 1 month) photospheric spectra of SN I. SN Ia are characterized by the presence of a deep absorption line produced by silicon, SiII $\lambda6355$. SN Ib, on the other hand, do not show such a line. Moderately strong He lines, especially He I$\lambda5876$, distinguish SN Ib from SN Ic at early times; in particular, SN Ic do not show this line. The late time (~ 5 months) optical spectra of SNe provide additional constraints: SN Ia show strong blends of forbidden emission lines of FeII and FeIII, while SNe of type Ib and Ic are instead dominated by emission lines of intermediate mass elements such as O and Ca, even though emission lines of Fe may also be present.

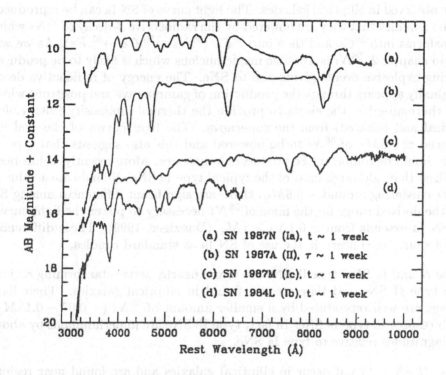

Figure 1.24. Spectra of supernovae showing the early time distinctions between the four major types and subtypes (II, Ia, Ib, Ic). The variable t indicates the time after the observed visual maximum. From Filippenko 1991, in "SN1987A and Other Supernovae", eds. I.J. Danziger and K. Kjär, p.343; reproduced here by kind permission of the European Southern Observatory (copy right 1991).

Supernovae of type II (SN II) are also subclassified into three different types according to their light curve shape: SN II-L (linear), SN II-BL (bright linear) and SN-P (plateau) (see Figure 1.25), and SN IIn, where the light curve can be strongly affected by the interaction between the SN ejecta and previous stellar winds. Late-time optical spectra of SN II are dominated by the strong H_α emission line and sometimes, though not always, forbidden lines of oxygen.

There are some fundamental observational facts concerning SNe of different type which help us to understand their possible progenitors.

- *Type Ia SNe*: the properties of type Ia SNe, such as the spectra, light curves and peak absolute magnitudes, indicate that most of them (the classical type Ia SNe) represent a very homogeneous class of objects, which makes them good candidates for standard candles and implies that their progenitors may also belong to a homogeneous class of objects.

They are found in galaxies of all morphological types and they are the only SNe observed in elliptical galaxies. The light curve of SN Ia can be reproduced if one assumes that they are powered by the radioactive β-decay of ^{56}Ni which transforms into ^{56}Co and then into ^{56}Fe ($^{56}Ni \rightarrow^{56} Co \rightarrow^{56} Fe$). As we will see in chapter 2, ^{56}Ni is a double magic nucleus which is likely to be produced during explosive events giving rise to SNe. The energy of radioactive decay originally appears through the production of gamma rays and positrons which are thermalized in the ejecta to provide the thermal luminosity (ultraviolet, optical and infrared) from the supernova. The light curves of classical SN Ia need $\simeq 0.6M_\odot$ of ^{56}Ni to be powered and this also suggests that type Ia SNe should be important Fe producers in galaxies. More recently, it has been realized that, although most of the typical type Ia SNe still indicate a value of ^{56}Ni clustering around $\sim 0.6M_\odot$, there are significant differences among SN Ia; the derived range for the mass of ^{56}Ni necessary to power the light curves of SN Ia extends from $\simeq 0.1$ to $\geq 1M_\odot$ (Danziger, 1999). These differences are a source of concern in the use of SN Ia as standard candles.

- *Type Ib and Ic SNe*: type Ib SNe are found nearby active star forming regions like type II SNe and they are not found in elliptical galaxies. Their light curves are well reproduced by a smaller amount of ^{56}Ni ($\sim 0.07 - 0.15M_\odot$) with respect to type Ia SNe. In fact, type Ib SNe are underluminous by about 2 magnitudes relative to type Ia SNe.

- *Type II SNe*: do not occur in elliptical galaxies and are found near regions of active star formation (HII regions). They are generally fainter than SN Ia and even fainter than SN Ib. The luminosity at the maximum of the light curve of SN Ia (generally assumed as standard candles) depends on the assumed Hubble constant and is $M_{B_{max}} = -19.79 \pm 0.12$ for $H_o = 46 \pm 10$ km sec^{-1} Mpc^{-1} (Tamman and Leibundgut, 1990), whereas for the luminosity at maximum of SN II one can adopt $< M_{B_{max}} >= -17.18$ and $\sigma_M = 1^m.2$ if $H_o = 50$ km sec^{-1} Mpc^{-1} (Tamman and Schroeder, 1990). The light curves of type II SNe are generally dominated by the radioactive decay of ^{56}Ni, well demonstrated by SN 1987A in LMC. It has been shown that one needs roughly $0.07M_\odot$ of ^{56}Ni to reproduce the light curve of SN 1987A. In Figure 1.26 is shown a compilation of data concerning the masses of ^{56}Ni derived for several type II SNe, including SN 1987A, and Ic SNe as functions of the mass of the stellar progenitor. From this figure one can see that apart from the peculiar case of SN 1997D, the derived mass of Ni is an increasing function of the initial stellar mass ranging from $\sim 0.07 M_\odot$ up to $\sim 0.7 M_\odot$. This kind of plot is extremely important for modelling the chemical evolution of galaxies since it imposes strong constraints on the Fe stellar production. The recent identification of a very energetic SN (SN 1998bw) associated with a γ-ray burst (GRB) and its modelling as the core-collapse of the carbon-oxygen core of a massive star ($\sim 40 M_\odot$), raises the question of the frequency of their occurrence particularly at early epochs as well as their chemical enrichment, since these objects seem to produce large amounts of oxygen and iron (see Figure 1.26).

In summary: the occurrence of SN Ia in ellipticals together with the absence of H in their spectra and the homogeneity of their spectra and light curves, with the exception of some subluminous type Ia SNe, argue for compact objects of nearly the same mass, originating from long lived stars, as possible progenitors of these SNe. The objects which possess all these characteristics are the white dwarfs (see chapter 2). On the other hand, the occurrence of type II, Ib and Ic SNe near star-forming regions and the presence of H in their spectra together with the different spectra and light curves argue in favor of short lived massive stars of varying mass as possible progenitors of these SNe (see chapter 2).

1.13.2   GALACTIC SUPERNOVA FREQUENCY

The first modern discussion of the galactic supernova frequency is by Opik (1953). He derived a rate of about one supernova per 300 years from the known supernova remnants. However, he realized that the true supernova rate must be considerably higher than this because of interstellar extinction, and because the historical record of galactic supernovae is incomplete. A comprehensive review of the galactic supernova rate can be found in van den Bergh and Tammann (1991). A compilation of SNe known to have occurred in the Galaxy during the last two millennia leads to one object during the first millennium and six during the second millennium. These supernovae have been classified according to the properties of their remnants. By considering only the SNe occurring within 4 kpc from the Sun, one can infer a rate slightly higher than 3 type II supernovae (deriving from massive stars, see also chapter 2) during the second millennium. From a rate of 3 SNe per millennium within 4 kpc and an assumed thickness of the galactic disk of $\sim$0.3 kpc, one finds a local supernova rate of $\sim 2 \cdot 10^{-4} SN kpc^{-3} yr^{-1}$. For a disk age of 10 Gyr it then follows that the total number of SNe exploded in the solar neighbourhood is $\sim 2 \cdot 10^6 kpc^{-3}$. The SNe with massive progenitors probably have a Pop I distribution and these stars follow an exponential radial density distribution with a scale length of 4 kpc extending from 3 to 15 kpc (Ratnatunga and van den Bergh, 1989). With such a

52

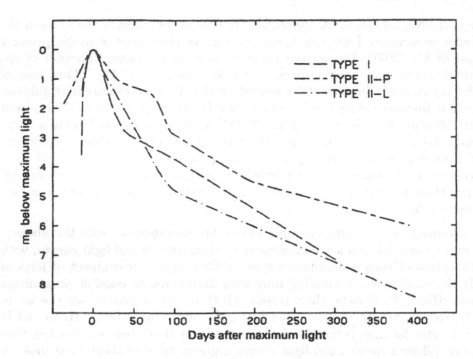

Figure 1.25. Typical light curves for type II-P and II-L SNe compared with the typical light curve of a type Ia SN. From Kirshner 1990, in "Supernovae", ed. A. Petschek, p. 59; reproduced here by kind permission of Springer Verlag (copy right 1990).

distribution, roughly 9% of all galactic OB stars will be located within 4 kpc from the Sun, then the true galactic rate would be 2-3/0.09=22-33 per millennium, namely $\sim$ 2-3 per century (the inclusion of type Ia SNe would increase this value by $20 - 25\%$).

From a review of all galactic luminosity indicators, van den Bergh (1988) concluded that the Milky Way has a blue luminosity $L_B = (2.3 \pm 0.6) \cdot 10^{10} L_{B_\odot}$. This value of the galactic luminosity when combined with a SN rate of 3 per century, yields a galactic rate of 1.3 SNU for SNe with massive progenitors (where SNU is the number of SNe per century per $10^{-10} L_{B_\odot}$).

1.13.3 SUPERNOVA FREQUENCY IN LATE TYPE GALAXIES

The galactic SN rate can also be estimated by comparison with external galaxies similar to the Milky Way. Since a galaxy luminosity depends on the Hubble constant H_o (in units of km sec^{-1} Mpc^{-1}), there is a factor h^2 (with $h = 50/H_o$) involved in the inferred SN frequencies. For a spiral galaxy similar to the Galaxy, namely for a galaxy between types Sab-Sb and types Sbc-

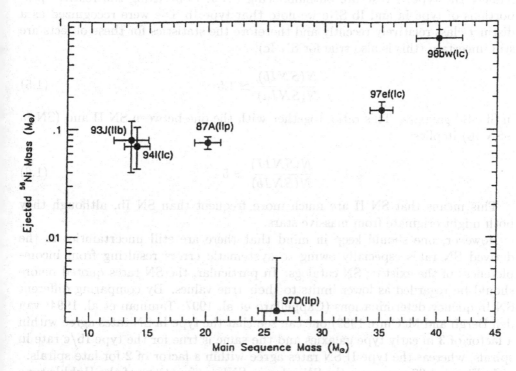

Figure 1.26. Ejected ^{56}Ni mass as a function of main-sequence mass, as estimated for SN II 1987A ; SN IIb 1993J; SN Ic 1994I; SN II 1997 D; SN Ic 1998bw; SN Ic 1997ef. From Iwamoto et al. 2000, Ap.J. 534, 660; reproduced here by kind permission of K. Iwamoto and the University of Chicago Press (copy right 2000).

Sd the inferred SN rates are: $(1.35 - 3.93)h^2 \cdot L_B(Gal)/10^{10}L_{B_\odot}$SNU for type II SNe, $(0.49)h^2 \cdot L_B(Gal)/10^{10}L_{B_\odot}$SNU for type Ia SNe and $(0.27 - 0.77)h^2 \cdot L_B(Gal)/10^{10}L_{B_\odot}$SNU for type Ib SNe.

The galactic SN rates are then obtained by multiplying the observed rates by the galactic blue luminosity, once a value for the Hubble constant has been chosen. By adopting this method, Cappellaro et al. (1997) obtained in a millennium 12 ± 6 SNe of type II, 4 ± 1 SNe of type Ia and 2 ± 1 SNe of type Ib, under the assumption of $L_B(Gal) = 2 \cdot 10^{10}L_{B_\odot}$ and $H_o = 75$ km sec^{-1} Mpc^{-1}.

The ratios among SN rates are very important for constraining the chemical evolution of galaxies. From supernova rates in late-type galaxies (spirals and magellanic irregulars) (see van den Bergh and Tammann, 1991) we know that :

$$\frac{N(SNII)}{N(SNIa + SNIb)} \simeq 3 \tag{1.4}$$

namely the type II SNe are outnumbering SN I. Concerning the relative proportions of type Ia and Ib SNe we note that type Ib SNe were recognized as a distinct class relatively recently and therefore the statistics for these objects are still uncertain (this is also true for SN Ic):

$$\frac{N(SNIb)}{N(SNIa)} \simeq 1.5 \qquad (1.5)$$

in Sbc-Sd galaxies. This ratio, together with the one between SN II and (SN Ia +SN Ib) implies:

$$\frac{N(SNII)}{N(SNIb)} \simeq 5 \qquad (1.6)$$

This means that SN II are much more frequent than SN Ib, although they both might originate from massive stars.

However, one should keep in mind that there are still uncertainties in the derived SN rates especially owing to systematic errors resulting from incompleteness of the existing SN catalogs. In particular, the SN rates quoted before should be regarded as lower limits to their true values. By comparing different SN frequency determinations (Cappellaro et al. 1997; Tamman et al. 1994; van den Bergh and McClure 1989) one can see that the type Ia SN rates agree within a factor of 3 in early type galaxies and the same is true for the type Ib/c rate in spirals, whereas the type II SN rates agree within a factor of 2 for late spirals.

In Figure 1.27 we report the SN rates in SNU as functions of the Hubble type of their host galaxies. It is worth noting that SN rates for galaxies of a given Hubble type correlate remarkably well with the blue luminosity of their parent galaxy, in the sense that the SN rate is higher for higher blue luminosities. In addition, the type II and Ib SN rates per unit luminosity increase from early to late galaxy types. In this case the early type galaxies are spirals since SN II and Ib are absent in elliptical and S0 galaxies. The type Ia rate per unit luminosity is higher in E-S0 galaxies than in spirals where it is found to be roughly constant. This again indicates that type II and Ib SNe, contrary to type Ia SNe, are connected with active star formation. Another interesting fact is that the expansion velocities of SN Ia appear to correlate with Hubble type, in the sense that supernovae in early-type galaxies tend to have lower expansion velocities than do those which occur in late-type galaxies. This might suggest that SN Ia with low expansion velocities may belong to an older less massive stellar population than do those with high expansion velocities.

1.13.4 COSMIC SUPERNOVA FREQUENCIES

It is interesting to compute the so-called cosmic SN rates, namely the SN rates weighted over the local blue luminosity function taking into account galaxies of all morphological types. This cosmic rate is important if one looks at the universe at high redshift where it is impossible to make a distinction among galaxies of different morphological type. Following Madau et al. (1998) one can weight the present time SN rates for different SN types by assigning a given percentage to galaxies of different morphological type contributing to the local blue luminosity function (28% to ellipticals, 32% to early spirals and 40% to late spirals). The resulting cosmic SN rates are: $SN_{Ia} = 0.14 \pm 0.06h^2$ SNU and $SN_{II} = 0.49 \pm 0.2h^2$ SNU. Searches for SNe at high redshift will provide a better insight into the cosmic rates at earlier times and will be very useful to test theories of SN progenitors (see chapter 2). At the moment, searches of distant SNe have provided the first measurement of the type Ia SN rate at $z = 0.4$, $SN_{Ia_{z=0.4}} = 0.21^{+0.17}_{-0.13}h^2$ SNU (Pain et al. 1996 and references therein).

1.14   NOVA RATES

Nova systems can be important in the study of the chemical evolution of galaxies since they have been proposed as important producers of ^7Li, ^{13}C and ^{15}N (Starrfield et al. 1978; Josè and Hernanz, 1998), even though the nucleosynthesis calculation for nova outbursts are still uncertain. The nature of novae will be discussed in chapter 2, while here we want only to review what is their observed rate in the Galaxy. In spite of the importance of an accurate determination of the galactic nova rate, the actual value is still poorly known. Early estimates of the nova rate (Allen 1954; Sharov 1972) indicated very high values of the order of 100-260 novae yr^{-1}. Later a value of 73 ± 24 novae yr^{-1} was suggested by Liller and Mayer (1987) at variance with estimates of nova rates in external spirals such as M31, where a nova rate of 29 ± 4 nova yr^{-1} was estimated by Capaccioli et al. (1989) and of 16 nova yr^{-1} by van den Bergh (1991). More recently, Della Valle and Livio (1994) suggested a galactic nova rate of $\simeq 20$ nova yr^{-1}, which is a factor 3-4 lower than the previous estimates for the Galaxy and more in agreement with nova rate estimates in external spirals.

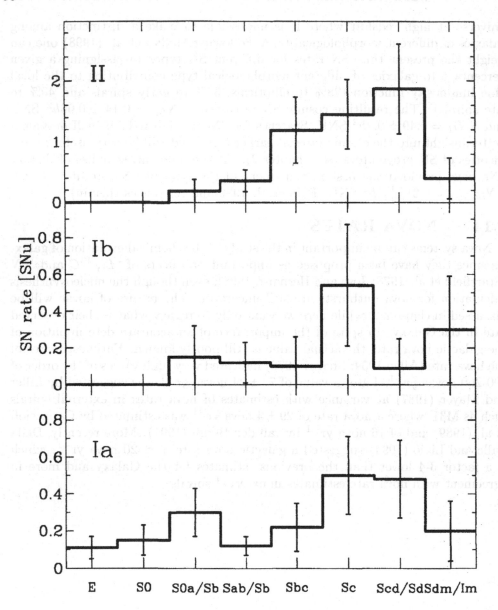

Figure 1.27. SN rates per unit blue luminosity (units of SNU, namely SNe $100^{-1}10^{-10}L_{B_\odot}$) for galaxies of different morphological types. The 1-standard deviation Poisson errors are also reported. When a single morphological type is considered, statistical errors become quite large, especially for SN Ib, whose discoveries are less frequent. The assumed Hubble constant is $H_o = 75$ km sec^{-1} Mpc^{-1}. From Cappellaro et al. 1993, A&A Vol. 273, 383; reproduced here by kind permission of Springer Verlag (copy right 1993).

Chapter 2

STELLAR EVOLUTION AND
NUCLEOSYNTHESIS

Here we address the phases of stellar evolution and nucleosynthesis which are more important in connection to chemical enrichment and galactic chemical evolution. In the following one should keep in mind that the stellar evolution results are influenced by the physics adopted in the models. In particular, two physical processes are important in determining the galactic chemical enrichment, the mass loss occurring during various phases of stellar evolution and the treatment of convection. Various types of mixing are expected to occur during the life of a star: convection in central cores, in shells, in external envelopes, overshooting, rotation induced turbulent diffusion. For the sake of clarity in the next sections we will define here what is defined as "overshooting". Usually the limit of a stellar convective core is set at the layer where the acceleration of the fluid elements is zero. However, the zero acceleration point does not coincide with the zero velocity point which should be the real limit of the convective core. Consideration of this effect in stellar evolution leads to overshooting (see Chiosi 1987). Most of the formulations for overshooting are based on the mixing-length theory and use the mean free path of convective elements expressed as $\alpha_{over} = 1/H_p$ (H_p is the pressure scale height and is a free parameter). Classical stellar evolution ignores the effects of many of those physical processes.

2.1 THE EVOLUTION OF SINGLE STARS

The evolutionary history of a star is the history of the gravitational contraction of a mass of gas under the effect of its own gravitational energy. When the density and temperature at its center reach the necessary conditions to ignite the nuclear fuel present at that moment, the gravitational contraction stops counterbalanced by the pressure generated by the nuclear energy production.

57

According to their initial mass, stars evolve quite differently and have very different lifetimes. In particular, the lifetime of a star of mass M is usually defined as the time spent in the Main Sequence (hereafter MS) phase, when hydrogen is turned into helium in the core of the star. In fact, the MS time represents 90% of the lifetime of each star. The stellar lifetimes increase with decreasing initial stellar mass, this is because the MS luminosity of the stars follows the law:

$$L_{MS} \propto M^4, \tag{2.1}$$

whereas the nuclear energy of stars during the MS phase goes roughly as:

$$E_N \propto M. \tag{2.2}$$

Therefore the MS lifetime of a star of mass M is:

$$t_{MS} \propto E_N/L_{MS} \propto M^{-3} \tag{2.3}$$

With increasing stellar mass, stars are able to ignite more and more nuclear fuels in their cores and stars with masses sufficiently large can eventually ignite all six main hydrostatic nuclear burnings (H-, He-, C-, Ne-, O- and Si-burning). After Si-burning, an inert Fe-core is formed and the nuclear evolution of the star is over. This is due to the fact that the binding energy per nucleon reaches a maximum for an atomic mass corresponding to that of iron, and the fission reactions are favored with respect to the fusion ones. This situation is typical for massive stars ($M > 10 - 12M_\odot$).

Before proceeding further we define the crucial ranges of masses and the masses delimiting such ranges. Following Iben and Renzini (1983) we divide the stars in:

Low mass stars, namely all the stars with masses $\leq M_{HeF}$, where M_{HeF} is the limiting mass for the formation of an electron- degenerate He-core immediately following the MS phase. These stars ignite He explosively, the so-called He-flash.

Intermediate mass stars, namely the stars in the range $M_{HeF} \leq M/M_\odot \leq M_{up}$, where M_{up} is defined as the limiting mass for the formation of an electron degenerate C-O core or the minimum initial mass for an off-center C-ignition proceeding up to central C-exhaustion. These stars ignite He non-degenerately but develop an electron degenerate C-O core following the exhaustion of He.

Massive stars, namely all the stars with masses $> M_{up}$, in other words the stars which do not develop an electron degenerate C-O core.

Other masses of interest for understanding stellar evolution and nucleosynthesis are:

- M_L, the limiting mass for H-ignition in the stellar core,
- M_w, the limiting mass for the formation of a C-O white dwarf,

- M_{SNII}, the limiting mass for the occurrence of a type II SN event. Stars with masses larger than M_{SNII} either become Wolf-Rayet stars and their explosion would not resemble that of a type II supernova or they would implode as black holes.

2.1.1   LOW MASS STARS

Here we summarize the main evolutionary stages of stars with masses $\leq M_{HeF}$. The precise value of M_{HeF} depends on the treatment of convection: the classic (no mass loss, no overshooting) value is $\simeq 2.2\ M_\odot$; with convective overshooting is lower, $\sim 1.85\ M_\odot$ (Maeder and Meynet, 1989), as we discuss in more detail later.

All the stars with masses $\leq M_L$, if they exist, never ignite H and their luminosity is simply given by the gravitational energy liberated in their slow contraction. They are called *brown dwarfs* and their existence is relevant to the nature of dark matter. The value of M_L again depends on the initial stellar metallicity, on the input physics of the stellar models and on atmosphere models; most models indicate $M_L \simeq 0.08 - 0.09 M_\odot$ (D'Antona and Mazzitelli, 1996). The brown dwarfs are also relevant to galactic chemical evolution in the sense that they subtract ISM gas from further processing.

Stars in the range $M_L - 0.5 M_\odot$ ignite hydrogen in the center but their He-cores become electron-degenerate before getting hot enough to burn helium. These stars will therefore end their lives as He-white dwarfs. These objects have lifetimes much larger than the age of the universe.

Stars with masses in the range $0.5 \leq M/M_\odot \leq M_{HeF}$ after the MS phase and Red Giant Branch phase (hereafter RGB, when H-burning occurs in a shell outside the He-core and the He-core is contracting), ignite He in an electron-degenerate core (He-flash), then become C-O white dwarfs after passing through the phases of Horizontal Branch (HB), Asymptotic Giant Branch (AGB) and PN phase (see Figure 2.1). During the RGB, AGB and PN phases the stars loose mass (a star with initial mass equal to $1 M_\odot$ becomes a C-O white dwarf of $\sim 0.6 M_\odot$). This mass loss is important since it is the only way in which these stars can restore their processed and unprocessed material into the ISM.

2.1.2   INTERMEDIATE MASS STARS

We summarize the main evolutionary phases of these stars in connection with the chemical enrichment of the ISM. The most important phase in this respect is the AGB phase during which dredge-up episodes are occurring and chemical elements freshly made in the core of the star may be transported into the external stellar layer and eventually be ejected into the ISM. These dregde-up episodes occur in between the so-called thermal pulses which are He-shell flashes experienced during the ascent of the AGB of low and intermediate mass stars. In the

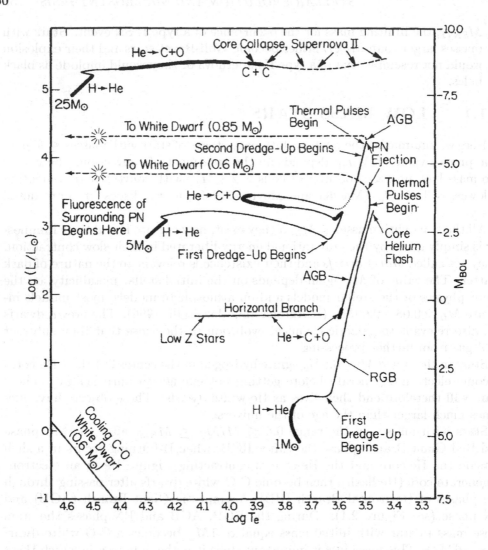

Figure 2.1. Tracks of stars of low ($1M_\odot$), intermediate ($5M_\odot$) and high ($25M_\odot$) mass in the theoretical H-R diagram. The places where the first and second dredge-up episodes occur are indicated, as are the places along the AGB where thermal pulses begin. The third dredge-up episode occurs during the thermal pulse phase. From Iben I.Jr. 1991, Ap.J. Suppl. Vol. 76, 55; reproduced by kind permission of I.Jr. Iben and the University of Chicago Press (copy right 1991).

He-burning convective shell, which develops during thermal pulses, neutron-rich isotopes are produced and in sufficiently massive AGB stars they are produced with the solar system *s-process* distribution. During the dredge-up episodes, occurring during the ascent of RGB and AGB and owing to the inward motion of the external convective zone, both fresh s-process elements and carbon are dredged-up into the convective envelope. Finally, in the AGB stars some of the dredged-up ^{12}C may be converted into ^{13}C and ^{14}N at the base of the envelope during the interpulse phase. This burning at the base of the convective envelope is known as *hot bottom burning* (Iben and Renzini 1983). If the dredged-up ^{12}C is the newly made one then the ^{13}C and ^{14}N have a *primary* origin. In particular, we define a primary element as an element synthesized starting from the original H and He, independently of metals. On the other hand, we define a *secondary* element as an element which is produced proportionally to the abundance of metals already present in the star at its birth and not made "in situ". ^{7}Li can also be produced and dredged-up into the convective envelope during the AGB phase. This element, in fact, is produced during H-burning only if ^{7}Be, produced through the reaction, $^{3}He(\alpha,\gamma)^{7}Be$, is rapidly transported by convection into regions where lower temperatures let it decay to ^{7}Li (see next section). In particular, if the temperature in not larger than $2.8 \cdot 10^{6}$K, ^{7}Li can survive, otherwise it reacts with an atom of H to give two α particles.

The value of M_{up} is very sensitive to the input physics in stellar models which can influence the size of the He-core. In particular, it is very sensitive to the treatment of convection. The classical stellar models with no convective overshooting predict $M_{up} = 7 - 9M_{\odot}$ (Becker and Iben 1979; 1980), as opposed to models with overshooting predicting $M_{up} = 5 - 6M_{\odot}$ (Bertelli et al. 1985; Maeder and Meynet 1989; Marigo et al. 1996). This is due to the fact that convective overshooting leads to larger He-core masses for a given initial mass. The mass M_{up} can vary also as a function of the stellar metallicity, as shown, for example, by Tornambè and Chieffi (1986). The metallicity in the star, in fact, influences both the efficiency of the H-burning shell and the size of the convective core determining the final size of the He-core. In particular, the relation between metallicity and M_{up} is not monotonic, as shown in Figure 2.2, where M_{up} first decreases with decreasing stellar metallicity until $Z \simeq 4 \cdot 10^{-4}$ and then it increases for lower metallicities but its value is always below $7M_{\odot}$.

Intermediate mass stars eject their processed and unprocessed material through stellar winds, present during RGB and AGB phases, and through the PN ejection before ending their lives as carbon-oxygen white dwarfs.

2.1.3 MASSIVE STARS

Stars which do not develop an electron-degenerate C-O core and ignite carbon non-degenerately are known as *massive stars*. Stars in the mass range $M_{up} < M/M_{\odot} < 10 - 12$ ignite C non-degenerately and those which have He-cores between 2.2 and 2.5 $M_{\odot}$ ignite O in a degenerate Ne-O core. The MS masses of these stars in classic models are 8-12 $M_{\odot}$, whereas in models with overshooting they are 6.6-10 $M_{\odot}$. These stars are believed to end their lives as SNe of type II, although their final state is still under debate. In particular, they are known as

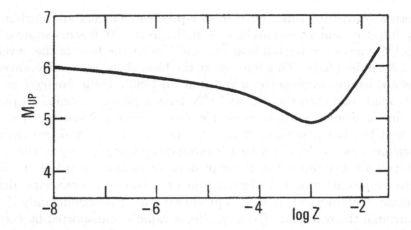

Figure 2.2. The value of M_{up} as a function of the metal content of the star. From Tornambè and Chieffi 1986, M.N.R.A.S. Vol. 220, 529; reproduced here by kind permission of Blackwell Science LTD (copy right 1986).

e-capture supernovae since their degenerate Ne-O-Mg core reaches the conditions for electron captures on nuclei (inverse β-decay). These e-captures induce the core to collapse into a neutron star, owing to the decrease in pressure of the electron degenerate gas, and during the collapse oxygen ignites transforming O to Si and finally to Fe. The collapse of the core eventually wins over the explosion since the star is not destroyed but it ejects its mantle leaving a neutron star of $\sim 1.3 M_{\odot}$ as a remnant (Nomoto 1984; 1987: Mayle and Wilson 1988).

Stars with initial masses in the range $10 - 12 \leq M/M_{\odot} \leq M_{WR}$, where M_{WR} is the limiting mass for the formation of a Wolf-Rayet star, namely a star which has undergone a huge mass loss and will end its life without the H-He envelope, evolve through all the six main hydrostatic nuclear burnings up to the formation of an Fe-core and explode as *iron-core collapse supernovae*. The mechanism for the explosion of these supernovae is not yet entirely clear owing to the difficulties encountered by modelers in getting these stars to explode. The basic idea is that there should be a core bounce giving rise to the ejection of the external mantle. This bounce is due to the fact that the Fe-core is contracting and when its density reaches the density of the nuclei of the atoms ($\simeq 2 \cdot 10^{14} g \, cm^{-3}$) the matter becomes incompressible and the collapse halts abruptly thus giving rise to the bounce. The bounce, in turn, produces pressure waves propagating towards the external parts. Such waves then become a shock wave with an energy equal to the gravitational energy of the collapse. The main problem for the explosion of these stars is that the energy produced by the explosion should be sufficient to heat the mantle of the star and eject it. In fact, the enormous energy in the shock wave after the core bounce is largely used to photo-disintegrate the outer core layers consisting of iron and is also lost through the neutrinos escaping from the hot core behind the shock during the neutronization process (see next section). Therefore, only a small fraction of the explosion energy is eventually available to heat the mantle, especially if the mass of the core is large. As a consequence, the shock wave, when it reaches the most external regions, has lost most of its energy and dies. Bethe and Wilson (1985) proposed a mechanism which allows the shock wave to rejuvenate and give rise to a *delayed explosion*. The rejuvenation is due to a small fraction of the neutrino energy- equal to the total gravitational binding energy of the neutron star of 10^{53} erg, which can be trapped in the very dense matter and give the shock wave the strength to eject the mantle. Type II SNe leave a neutron star or a black hole after the explosion.

Stars in the mass range $M_{WR} \leq M/M_\odot \leq 100$ can end up as supernovae but not of type II since they are missing their H-envelope, owing to the mass loss, whereas SN II are characterized by H lines present in their spectra. On one hand, if the collapsed proto-neutron star exceeds the maximum neutron star mass (which is still uncertain, $1.4 - 2.0M_\odot$) a collapse to a black hole occurs and prevents a SN explosion with the mechanism described above. On the other hand, there are recent indications that the accretion onto such a black hole can cause γ-ray bursts and eject the outer mantle, causing what has been called a *hypernova* event (McFayden and Woosley 1999). It has been suggested that these stars might be the progenitors of type Ib SNe which show a lack of H in their spectra (see the following sections). Unfortunately, we do not know exactly the limiting mass for the formation of Wolf-Rayet stars, since this depends on the assumed mass loss rate which, in turn, depends on the stellar metal content (radiation pressure mechanism). For a solar metallicity $M_{WR} \geq 40 \ M_\odot$ (Maeder, 1992).

Stars with masses $M > 100M_\odot$, following the exhaustion of He in the central regions, contract and proceed directly to O-burning. They undergo pair-creation events during O-burning. The pair-creation events occur because a large portion of the energy from gravitational contraction goes, at $T \sim 2 \cdot 10^9$ K, into creation of pairs (e^+, e^-). This reduces the polytropic index Γ (the ratio of the specific heats in a photon gas) below $4/3$ and induces instability in the star. The pair-creation instability (Fowler and Hoyle, 1964) may lead to violent pulsational instability with ejection of the external layers and later to iron core collapse ($35 \leq M_\alpha/M_\odot \leq 60$, where M_α is the mass of the He-core), or to complete explosion ($60 \leq M_\alpha/M_\odot \leq 110$), or to collapse to black holes directly after He-exhaustion ($M_\alpha \geq 110M_\odot$) (Woosley, 1987).

Supermassive objects are those which collapse directly to black holes or suffer total disruption due to explosive H-burning. The minimum initial mass for these objects is $M \sim 410 \ M_\odot$ (Appenzeller and Tscharnuter, 1973). Masses $M > 7.5 \cdot 10^5 \ M_\odot$ would end up as massive black holes (Appenzeller and Fricke, 1972). Stars in the range $410 < M/M_\odot < 7.5 \cdot 10^5$ suffer total disruption during H-burning, but the results are very sensitive to the initial stellar metal content. In particular, the explosion is possible only for solar metal content.

2.2 NUCLEOSYNTHESIS

2.2.1 THE H-BURNING

Here we briefly describe the main hydrostatic burnings in stars which proceed from H via He, ^{12}C, ^{20}Ne, ^{16}O and ^{28}Si to ^{56}Fe recalling that the *atomic number* of a nucleus is the number of its protons (Z) and the *mass number* of a nucleus is the quantity $A = Z + N$, which is the sum of its protons and neutrons.

The first important nuclear reactions in stars are those which transform H into He. There are two principal sequences of reactions which result in the conversion of four protons into a nucleus of ^4He and they are: (a) the proton-proton (p-p) chain and (b) the CN cycle.

a) The reactions of the main branch of the p-p chain are:

$$\text{1. } ^1H + {}^1H \rightarrow^2 D + e^+ + \nu_e \qquad (2.4)$$

$$\text{2. } ^2D + {}^1H \rightarrow^3 He + \gamma \qquad (2.5)$$

$$\text{3. } ^3He + {}^3He \rightarrow^4 He +^1 H +^1 H \qquad (2.6)$$

The nucleus of deuterium, composed by a proton and a neutron, is always destroyed, also the preexisting one, at temperatures of $\simeq 10^6 K$. The total energy production in the sequence is 26.7 MeV or $4.27 \cdot 10^{-5}$ erg. The neutrinos escape immediately from the star carrying part of this energy.

The second branch of the p-p chain occurs when ^3He, produced by the reaction 2, reacts with ^4He. The sequence is then:

$$\text{3'. } ^3He + {}^4He \rightarrow^7 Be + \gamma \qquad (2.7)$$

$$\text{4'. } ^7Be + e^- \rightarrow^7 Li + \nu_e + \gamma \qquad (2.8)$$

$$\text{5'. } ^7Li + {}^1H \rightarrow^4 He +^4 He \qquad (2.9)$$

The last reaction results in the destruction of the ^7Li, also the one initially present in the star, when the temperature is larger than $2.8 \cdot 10^6 K$.

The third branch of the p-p chain occurs if a proton instead of an electron is captured by the ^7Be:

$$\text{4''. } ^7Be + {}^1H \rightarrow^8 B + \gamma \qquad (2.10)$$

$$5''. \quad {}^8B \to e^+ + \nu_e + {}^8Be \to {}^4He + {}^4He + e^+ + \nu_e \tag{2.11}$$

An approximate expression for the energy production in solar conditions and at a temperature of $\simeq 1.5 \cdot 10^7 K$ is :

$$\epsilon_{pp} = 0.45\rho X^2 (T_6/15)^{3.95} \tag{2.12}$$

This quantity is expressed in units of [erg g^{-1} sec^{-1}], X is the H mass fraction, ρ is the density and T_6 is the temperature in units of $10^6 K$.

b) The CN cycle becomes more important than the p-p chain only at temperatures $T > 2 \cdot 10^7 K$. The reactions of the CN cycle are:

$$1. \quad {}^{12}C + {}^1H \to {}^{13}N + \gamma \tag{2.13}$$

$$2. \quad {}^{13}N \to {}^{13}C + e^+ + \nu_e \tag{2.14}$$

$$3. \quad {}^{13}C + {}^1H \to {}^{14}N + \gamma \tag{2.15}$$

$$4. \quad {}^{14}N + {}^1H \to {}^{15}O + \gamma \tag{2.16}$$

$$5. \quad {}^{15}O \to {}^{15}N + e^+ + \nu_e \tag{2.17}$$

$$6. \quad {}^{15}N + {}^1H \to {}^{12}C + {}^4He \tag{2.18}$$

The net effect of this series of reactions is the conversion of four protons into a nucleus of He and the production of two positrons (which annihilate) and two neutrinos.

A secondary branch of the CN cycle operates when ^{16}O is present:

$$6'. \quad {}^{15}N + {}^1H \to {}^{16}O + \gamma \tag{2.19}$$

$$7'. \quad {}^{16}O + {}^1H \to {}^{17}F + \gamma \tag{2.20}$$

$$8'. \quad {}^{17}F \to {}^{17}O + e^+ + \nu_e \tag{2.21}$$

$$9'. \quad {}^{17}O + {}^1H \to {}^{14}N + {}^4He \tag{2.22}$$

The CNO nuclei are acting as catalysts but at the completion of the reaction chain their relative abundances have changed whereas the total CNO abundance has remained constant. In normal stellar conditions the rate of the first reaction is 100 times higher than the rate of the fourth reaction which uses ^{14}N. All the other reactions are much faster and therefore, when a steady state equilibrium is reached, where each CNO nucleus is produced as rapidly as it is destroyed, the majority of the ^{12}C and the other nuclei have been converted into ^{14}N. The oxygen is generally produced less rapidly than the nitrogen is destroyed with the consequence that when this branch of the CNO cycle reaches the equilibrium conditions, the O/N ratio is reduced. Ultimately, the global effect of the CNO cycle is to convert 98% of the CNO isotopes into ^{14}N. Therefore, the N produced in this way is a *secondary* element as long as the CNO nuclei were originally present in the star at its birth and not produced *in situ*. In such a case the N produced is a *primary* element.

An expression for the energy produced per unit mass and unit time by the CNO cycle and valid for T_6 in the range 22–28 is:

$$\epsilon_{CNO} = 2.16 \cdot 10^4 \rho X X_{CNO}(T_6/25)^{16.7} \tag{2.23}$$

expressed in [erg g^{-1} sec^{-1}], where X is the H mass fraction and X_{CNO} is the mass fraction of CNO isotopes.

2.2.2   HE-BURNING AND BEYOND

The synthesis of elements heavier than He is not an easy task mainly because there do not exist stable isotopes with mass numbers A=5 and A=8. If two nuclei of ^4He merge to form ^8Be, this element decays immediately into two nuclei of ^4He. The He-burning should occur by means of a three particle reaction, the 3α reaction:

$$^4He +^4 He \rightarrow \leftarrow ^8 Be \tag{2.24}$$

$$^8Be +^4 He \rightarrow (^{12}C)^* + \gamma + \gamma \tag{2.25}$$

The 3α reaction can occur since, although ^8Be is unstable, the reaction (2.24) is in chemical equilibrium supporting at any time an existing (although very small) abundance of ^8Be. Salpeter showed that, under particular conditions, namely for $T = 10^8 K$ and $\rho = 10^5 g\,cm^{-3}$, there is a large enough He abundance to allow a third He nucleus to react with the ^8Be. The 3α reaction is a resonant reaction, namely it creates a nucleus of ^{12}C in an excited nuclear state (a resonance is normally indicated by $(^{12}C)^*$) and this increases the probability for its occurrence. The energy produced in this cycle is 7.274 MeV.

An approximate expression for the $\epsilon_{3\alpha}$ at $T = 10^8 K$ (where T_8 is the temperature expressed in units of 10^8 K) is:

$$\epsilon_{3\alpha} = 4.4 \cdot 10^{-8} \rho^2 Y^3 T_8^{40} \tag{2.26}$$

expressed in [erg g^{-1} sec^{-1}], where Y represents the He mass fraction.

Other reactions which occur during He-burning but at temperatures slightly higher than the 3α reaction are:

$$^{12}C +^4 He \rightarrow^{16} O + \gamma \qquad (2.27)$$

$$^{16}O +^4 He \rightarrow^{20} Ne + \gamma \qquad (2.28)$$

$$(^{20}Ne +^4 He \rightarrow^{24} Mg + \gamma) \qquad (2.29)$$

$$(^{24}Mg +^4 He \rightarrow^{28} Si + \gamma) \qquad (2.30)$$

Of these reactions only the first two are important so that at the end of He-burning we expect a mixture of ^{12}C and ^{16}O together with some ^{20}Ne, although the majority of ^{20}Ne is produced during C-burning.

An important fact is that during the He-burning there is capture of α particles on ^{14}N nuclei synthesized during the previous CNO cycle; in particular it is interesting to recall the following chain of reactions which leads to the production of neutrons that can then be captured by Fe seed nuclei thus producing s-process elements:

$$^{14}N(\alpha, \gamma)^{18} F(\beta^+)^{18}O(\alpha, \gamma)^{22} Ne(\alpha, n)^{25} Mg \qquad (2.31)$$

Such a chain is possible if before the end of the He-burning temperatures of the order of $2 \cdot 10^8 K$ are reached.

When the temperature reaches the value of $7 - 8 \cdot 10^8 K$, the first reaction is:

$$^{12}C +^{12} C \rightarrow (^{24}Mg)^* \rightarrow^{20} Ne + \alpha \qquad (2.32)$$

the probabilty of occurrence of this reaction is 50% and the energy released is 4.6 Mev.

$$^{12}C +^{12} C \rightarrow (^{24}Mg)^* \rightarrow^{23} Na + p \qquad (2.33)$$

with a probabilty of 50% and energy of 2.2 Mev,

$$^{12}C +^{12} C \rightarrow (^{24}Mg)^* \rightarrow^{23} Mg + n \qquad (2.34)$$

which is rare and with negative energy,

$$^{12}C +^{12} C \rightarrow (^{24}Mg)^* \rightarrow^{24} Mg + \gamma \qquad (2.35)$$

which is negligible and with energy of 13.9 MeV.

The second of these reactions leads again to the ^{20}Ne since the ^{23}Na reacts with the free protons in the following way:

$$^{23}Na(p, \alpha)^{20}Ne \tag{2.36}$$

At even higher temperatures the photodisintegration of Ne starts:

$$^{20}Ne + \gamma \rightarrow^{16} O + \alpha \tag{2.37}$$

which is followed by:

$$^{20}Ne + \alpha \rightarrow^{24} Mg + \gamma \tag{2.38}$$

$$^{24}Mg + \alpha \rightarrow^{28} Si + \gamma \tag{2.39}$$

When the temperature reaches the value T$\simeq 1.4 \cdot 10^9 K$ one has:

$$^{16}O +^{16} O \rightarrow (^{32}S)^* \rightarrow^{28} Si + \alpha \tag{2.40}$$

with an energy of 9.6 MeV,

$$^{16}O +^{16} O \rightarrow (^{32}S)^* \rightarrow^{31} P + p \tag{2.41}$$

with an energy of 7.7 MeV,

$$^{16}O +^{16} O \rightarrow (^{32}S)^* \rightarrow^{31} S + n \tag{2.42}$$

with an energy of 1.5 Mev,

$$^{16}O +^{16} O \rightarrow (^{32}S)^* \rightarrow^{32} S + \gamma \tag{2.43}$$

which is negligible and produce an energy of 16.5 MeV. Therefore, the main products of O-burning are ^{28}Si and ^{31}P, although the majority of ^{31}P is destroyed by (p,α) reactions and becomes ^{28}Si.

At T=$3 - 5 \cdot 10^9 K$ the creation of α particles from disintegration and the capture of α particles determines a thermodynamic equilibrium, the so-called nuclear statistic equilibrium (NSE), during which the formation of the nuclei with highest binding energy is favored, thus starting from nuclei of ^{24}Mg and ^{28}Si we have the formation of ^{36}Ar, ^{40}Ca, ^{44}Sc, ^{48}Ti, ^{52}Cr but mainly ^{56}Ni. In this way much ^{56}Ni is created and this will decay through two β^+ decays into ^{56}Co and ^{56}Fe:

$$^{56}Ni \rightarrow (\beta^+) \rightarrow^{56} Co \rightarrow (\beta^+) \rightarrow^{56} Fe \tag{2.44}$$

We note that β decays can be represented by the following reactions:

$$\beta^-[n \rightarrow p + e^- + \overline{\nu_e}] \tag{2.45}$$

$$\beta^+[p \rightarrow n + e^+ + \nu_e]. \tag{2.46}$$

The reaction (2.46) is possible only inside the nucleus, whereas it is not energetically possible for a free proton. Other weak interactions are related to (2.45) and (2.46) by taking particles from one side of the reactions and inserting them as their antiparticles on the other side, e.g.:

$$p + e^- \rightarrow n + \nu_e \tag{2.47}$$

$$n + e^+ \rightarrow p + \tilde{\nu}_e \tag{2.48}$$

$$\tilde{\nu}_e + p \rightarrow n + e^+ \tag{2.49}$$

$$\nu_e + n \rightarrow p + e^- \tag{2.50}$$

Reaction (2.47) is also known as the *neutronization process* which acts during the formation of a neutron star during a supernova explosion. It is worth recalling here that with the nucleus ^{56}Fe the binding energy per nucleon reaches a maximum and then decreases towards heavier nuclei, so that any further fusion reaction is not favored in a NSE. This is why stars when they reach the point of having an iron core they are "out" of fuel and contract towards their final state of neutron star or black hole after the ejection of their mantle (type II SN explosion).

2.2.3   THE ELEMENTS BEYOND THE IRON PEAK

The elements beyond the iron peak $(A > 60)$ can be formed through slow and/or rapid neutron capture on seed nuclei, mainly ^{56}Fe, since in comparison with a mixture of ^{56}Fe and free neutrons the total binding energy of heavier nuclei is still higher. These elements are known as s and r-*elements* according to the slow or rapid neutron capture relative to the timescale of β decay ($\tau_\beta \simeq 10^5 - 10^7$ sec). The process of slow neutron capture leads to the formation of very heavy nuclei. This process generally acts during the quiescent He-burning either in the stellar core or in the envelope. Since the neutrons made during quiescent nuclear burnings are produced in a moderate flux, one expects that the neutron capture process would be slow relative to the β decay.

The neutrons arise either from reaction (2.31) or from an alternative source even stronger than the previous one:

$$^{13}C(\alpha, n)^{16}O \tag{2.51}$$

which can be created in shell He-burning when protons are mixed into He-layers leading to:

$$^{12}C(p, \gamma)^{13}N(\beta^+)^{13}C(\alpha, n)^{16}O. \tag{2.52}$$

Then through a series of slow neutron captures followed by β- decays, elements up to Pb and Bi can be formed. However, not all elements heavier than $A > 60$ can be explained through *s-processes* since there are some elements in the Z-N plot which can be explained only if they had at least two neutron captures before β decaying. These elements should derive from *r-processes*, namely after a rapid capture of neutrons relative to the β decay. This requires a strong neutron flux which can be produced only during explosive events; one option is the explosive He-burning where the same reactions occurring in the quiescent He-burning are present on short timescales. On the other hand, these cannot be the only sources of *r-process* elements because they are too weak, as has been shown in the past. The neutron-rich environment close to the neutron star surface in SNe of type II is required to explain the *r-process* elements up to U and Th.

2.2.4   EXPLOSIVE NUCLEOSYNTHESIS

When the stellar iron core eventually collapses, due to photo-disintegration and electron capture, at the core bounce a shock wave is generated. Typically this shock wave stalls due to energy loss from photo-disintegrations of Fe-group nuclei in the resulting high temperatures. Some other mechanism is therefore needed to revive this shock wave (ν-heating from the hot neutron star, rotation, magnetic fields). Then, the shock propagates through the Si-O layers and explosive nucleosynthesis takes place in and behind the shock wave. The passage of the shock wave compresses and heats the stellar material raising the temperature up to the burning conditions. The fuels for explosive nucleosynthesis consist mostly of α-elements, such as ^{12}C, ^{16}O, ^{20}Ne, ^{24}Mg and ^{28}Si, namely all the elements synthesized by assembling α-particles. Since the timescale of explosive processing is very short (from a fraction of a second to several seconds), only few β decays take place during the explosive nucleosynthesis event thus favoring the formation of nuclei with Z=N, although the spread of nuclei around the N=Z line can be large. According to the temperature, intermediate to heavy nuclei are formed in explosive nucleosynthesis events. It is worth noting that while the main hydrostatic burnings proceed in the following sequence: H-, He-, C-, Ne-, O-and Si-burning, in explosive nucleosynthesis they proceed in the opposite direction, from Si to H-burning, since the shock moves from the center to the surface of the star.

At temperatures of the order of $4 - 5 \cdot 10^9$K, explosive Si-burning takes place and it can be either complete or incomplete with or without α-rich freeze-out. The α-rich freeze-out occurs generally at low densities when the 3α reaction is not fast enough to keep the He abundances in equilibrium with heavier nuclei during the fast expansion and cooling in the explosive events.

For $T > 5 \cdot 10^9$K the Si-burning is complete and leads to the formation of only ^{56}Ni (Fe-group nuclei). At such high temperatures, in fact, all the Coulomb barriers can be overcome and the abundances are in a full NSE for high densities and temperatures. For peak densities below $\rho \sim 10^8 g\, cm^{-3}$ an α-rich freeze out occurs.

Incomplete Si-burning occurs at lower temperatures ($T \sim 4 - 5 \cdot 10^9$K). In the Si-shell, besides the dominant fuel made of ^{28}Si and ^{32}S, the other most abundant nuclei are ^{36}Ar and ^{40}Ca. Starting from this material the incomplete Si-burning leads to the formation of iron-peak elements with ^{56}Ni and ^{54}Fe as dominant abundances in the Fe-group, plus smaller amounts of ^{52}Fe, ^{58}Ni, ^{55}Co and ^{57}Ni.

When the temperature is in excess of $T \sim 3.3 \cdot 10^9$K, explosive O-burning becomes active and leads to the production of mainly ^{28}Si, ^{32}S, ^{36}Ar, ^{40}Ca, ^{34}S and ^{38}Ar. In zones with temperatures close to $T = 4 \cdot 10^9$K some Fe-group nuclei are also produced (^{56}Ni and ^{54}Fe).

In the explosive Ne-, C- and He-burning the main products are ^{16}O, ^{24}Mg and ^{28}Si synthesized via the same sequence of reactions as in the hydrostatic case. Besides the main products, explosive Ne-burning supplies also substantial amounts of ^{27}Al, ^{29}Si, ^{32}S, ^{30}Si and ^{31}P, while explosive C-burning contributes in addition to ^{20}Ne, ^{23}Na, ^{24}Mg, ^{25}Mg and ^{26}Mg. Explosive He and H-burning essentially do not take place in low density environments like core collapse supernovae of massive stars, because the timescales of these processes are too long for the temperatures attained. Only minor side reactions occur which can produce nuclei with very low abundances. In high density environments explosive He-burning can occur and even create temperatures high enough that its products, ^{12}C and ^{16}O, ignite as well, changing finally into explosive Si-burning which leads again to the creation of ^{56}Ni. Explosive H-burning does take place in nova explosions and X-ray bursts (and possibly γ-ray bursts) occurring in close binary systems when H-rich material is transferred into a compact companion (a white dwarf or a neutron star). Explosive H-burning occurs through the CNO-cycle and is unstable to flash because of the electron-degenerate conditions where the pressure is not linked to the temperature as in a perfect gas but only to the density of the electrons. This leads to the occurrence of a thermonuclear runaway causing explosive ejection. Nuclei such as ^{15}N, ^{13}C and ^7Li can be produced during explosive H-burning, but in X-ray bursts temperatures can reach values even beyond 10^9 K and create nuclei beyond Fe via the *p-process* (rapid proton captures and β^+-decays).

2.2.5   THERMONUCLEAR SUPERNOVAE IN BINARY SYSTEMS

In recent years a great deal of attention has been paid to the study of the outcome of thermonuclear explosions of C-O white dwarfs (WDs) in binary systems since they are considered the most likely progenitors of supernovae of type Ia when they are so strong as to destroy the system, and of novae when the explosion involves only the accreted external layers of the white dwarf and the net effect of the explosion is the loss of some material. The original idea that supernovae Ia are triggered by the thermonuclear runaway of carbon burning in electron-degenerate cores is due to Hoyle and Fowler (1960). White dwarfs composed of carbon and oxygen represent the final state of intermediate mass stars. If isolated, they undergo cooling and eventually become dark matter but if they are in close binary systems they evolve in a completely different way. In a binary system, when the companion star evolves and expands, matter is tranferred onto the white dwarf which is rejuvenated and, in some cases undergoes a thermonuclear explosion which gives rise to a supernova Ia. Theoretically, two main evolutionary scenarios have been considered: i) merging due to loss of angular momentum caused by gravitational wave radiation of double C-O white dwarfs with a total mass exceeding the Chandrasekhar mass for a C-O WD ($M_{Ch} \sim 1.44 M_\odot$), and ii) accretion of H or He by mass transfer from the companion at a relatively high rate. The accreting white dwarf undergoes H or He burning near the surface which increases or decreases the white dwarf mass, according to the rate of mass accretion. The mass of the white dwarf can decrease because its matter can be lost as the result of the outburst which follows the explosive ignition of H, as in nova systems. On the other hand, the C-O material can increase due to quiet burning. In particular, Chandrasekhar white dwarfs can be formed with relatively high accretion rates, $\dot{M} \sim 10^{-8}$ to $10^{-6} M_\odot$ yr^{-1}, because of the relatively small ejection after H or He burning. In Figure 2.3 is shown the situation for the accretion of He-rich material onto a C-O WD. According to the different mass accretion rates and initial masses of the white dwarf different types of explosion will occur. The most likely companion of the white dwarf in this scenario are stars which become red giants and therefore are low and intermediate mass stars.

In both cases (i) and ii) the white dwarf explodes when it reaches M_{Ch}, as is often described in the literature. In the current literature, however, it is not very clear how in detail a C-O white dwarf explodes by a thermonuclear runaway when it approaches the Chandrasekhar mass limit. This limit is the limiting mass for the stability of an electron degenerate C-O object, in the sense that when the star approaches this limit the pressure exerted by the degenerate electrons is no more sufficient to counterbalance the gravity and the star starts collapsing. During the collapse the density increases and when it reaches $\sim 10^9 g\, cm^{-3}$ carbon ignites in the central region, and since the ignition occurs in a highly degenerate object a thermonuclear runaway sets in, because under these conditions the pressure of the gas depends only on its density and not on its temperature. Carbon ignites because an object approaching M_{Ch} has reached the limiting mass necessary for C-ignition ($\sim 1.39 M_\odot$). After C-ignition in the center, the explosive nuclear flame propagates outward and the outcome of this explosion depends on how the nuclear flames and shock waves propagate inside the white dwarf. The flame front is subject to various instabilities such

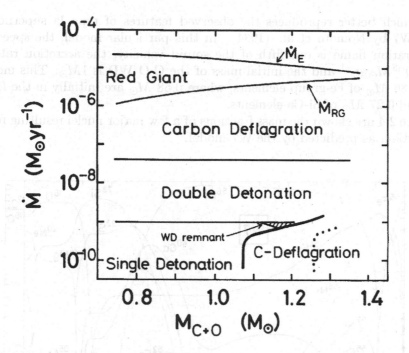

Figure 2.3.   Models of supernovae induced by accretion of He onto C-O WDs in the plane (M_{He}, M_{C+O}), where M_{C+O} denotes the initial mass of the WD. $\dot{M}_{RG}$ represents the rate of accretion of He onto the He-core in the red giant phase due to the H-burning shell and $\dot{M}_E$ is the Edddington accretion rate which is the limit above which the star is no more stable. Double detonation means that both He and C are ignited. From Nomoto et al. 1984, Ap.J. Vol. 286, 644; reproduced by kind permission of K. Nomoto and the University of Chicago Press (copy right 1984).

as Rayleigh-Taylor and Kelvin-Helmoltz types. In particular, the flame speed depends on the development of these instabilities and the resulting turbulence. According to the flame speed the explosion can be a C-detonation (supersonic speed) or a C-deflagration (subsonic speed) and the resulting nucleosynthesis is different in the two cases. In general, a C-detonation incinerates the whole object into ^{56}Ni whereas the C-deflagration produces a certain amount of ^{56}Ni plus intermediate mass elements.

Only particular models can account for the basic features observed in typical SN Ia, namely an explosion energy of $\sim 10^{51}$ erg, the synthesis of a large amount of ^{56}Ni (from 0.4 to 1.0 $M_\odot$) as well as the production of a substantial amount of intermediate mass elements at expansion velocities of ~ 10000 km sec^{-1} near the maximum brightness of the SN explosion. C-detonation is ruled out because it is not able to explain the presence of the intermediate mass elements, thus only C-deflagration is considered. Among the different models including C-deflagration,

74

the one which better reproduces the observed features of type Ia supernovae
is model W7 by Nomoto et al. (1984). In this particular model, the speed of
the deflagration flame is one fifth of the sound velocity, the accretion rate is
$\dot{M} = 4 \cdot 10^{-8} M_\odot yr^{-1}$ and the initial mass of the C-O WD is $1 M_\odot$. This model
predicts 0.86 $M_\odot$ of Fe-group elements, where 0.58 $M_\odot$ are initially in the form
of ^{56}Ni and 0.27 $M_\odot$ of Si-Ca elements.

In Figure 2.4 are shown the mass fractions of a few major nuclei resulting from
C-deflagration, as predicted by the W7 model.

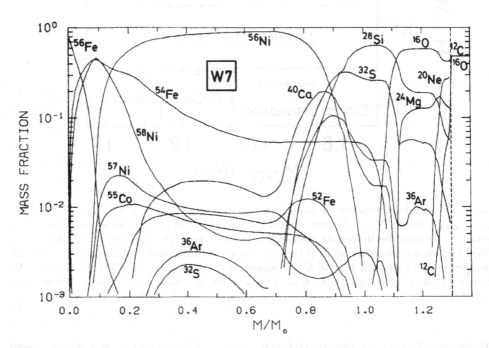

Figure 2.4. Mass fractions of few major nuclei resulting from C-deflagration as predicted by
model W7. From Nomoto et al. 1984, Ap.J. Vol. 286, 644; reproduced by kind permission of
K. Nomoto and the University of Chicago Press (copy right 1984).

More recently, it has been suggested that also sub-Chandrasekhar white dwarfs
can give rise to a SN explosion as the possible outcome of accretion with mass
accretion rates in the range $4 \cdot 10^{-8} - 10^{-9} M_\odot yr^{-1}$. In this case, in fact, He
ignites at high densities and the He-shell flash is strong enough to initiate an
off-center He-detonation. However, light curves and spectra predicted by these
sub-Chandrasekhar models for type Ia SNe seem not to agree with observations.

In all of these proposed models there are still many uncertain parameters such
as the velocity of the deflagration flame, the detonation-deflagration transition
density and the central ignition density, which prevent us from establishing a
clear progenitor model for supernovae of type Ia.

2.2.6 SUMMARY OF NUCLEAR BURNINGS

The most important hydrostatic nuclear burnings in stars can be summarized as follows:

- *H-burning*: the two reaction sequences are: the p-p chain which is initiated by $H(p, e^+)D$ and the CNO cycle, a sequence of (p, γ) and (p, α) reactions on C, N and O isotopes and subsequent β decays. In the normal MS stars, the CNO isotopes are all transformed into ^{14}N, due to the fact that the reaction $^{14}N(p, \gamma)^{15}O$ is the slowest reaction in the cycle.

- *He-burning*: the main reactions are the triple-α reaction $^4He(2\alpha, \gamma)^{12}C$ and $^{12}C(\alpha, \gamma)^{16}O$. It is worth noting that the uncertainties in the rate of the $^{12}C(\alpha, \gamma)^{16}O$ reaction cause a considerable uncertainty in the nucleosynthesis production and final fate of massive stars.

- *C-burning* : $^{12}C(^{12}C, \alpha)^{20}Ne$ and $^{12}C(^{12}C, p)^{23}Na$. However most of the ^{23}Na nuclei will react with the free protons via the $^{23}Na(p, \alpha)^{20}Ne$.

- *Ne-burning* : $^{20}Ne(\gamma, \alpha)^{16}O$, $^{20}Ne(\alpha, \gamma)^{24}Mg$ and $^{24}Mg(\alpha, \gamma)^{28}Si$.

- *O-burning* : $^{16}O(^{16}O, \alpha)^{28}Si$, $^{16}O(^{16}O, p)^{31}P$ and $^{16}O(^{16}O, n)^{31}S(\beta^+)^{31}P$. In analogy with C-burning most of the ^{31}P is destroyed by a (p, α) reaction to ^{28}Si.

- *Si-burning* : Si-burning is initiated by photodisintegration reactions which then provide the particles for capture reactions at $T \geq 3 \cdot 10^9$ K and it ends in an equilibrium abundance distribution around ^{56}Ni.

The explosive nuclear burnings in massive stars proceed in the following way:

- *explosive Si-burning*: a) complete, all Si is burned. In a normal freeze-out it is transformed into Fe-group elements, mostly into ^{56}Ni or b) incomplete, Fe-group elements, mostly ^{56}Ni and ^{54}Fe combined with ^{28}Si, ^{32}S, ^{36}Ar, ^{40}Ca,

- *explosive O-burning*: mostly ^{28}Si, ^{32}S, ^{34}S, ^{36}Ar, ^{38}Ar, ^{40}Ca plus traces of ^{56}Ni and ^{54}Fe,

- *explosive Ne-burning*: mostly ^{16}O, ^{24}Mg and ^{28}Si plus others,

- *explosive C-burning*: produces ^{20}Ne, ^{23}Na and Mg isotopes,

- *explosive He-burning*: negligible in SN II but possible in high density environments, leading finally to explosive Si-burning,

- *explosive H-burning* : ^{15}N, ^{13}C and ^7Li, in high density environments leading to the p-process.

Figure 2.5 shows the mass fractions of major nuclei resulting from hydrostatic burning stages prior to core collapse calculated for an initially $20M_\odot$ star. In Figure 2.6 is illustrated the situation of the same star after the occurrence of the explosive nucleosynthesis processes. From this figure is evident that the most important changes are due to Si-, O- and Ne-explosive-burning.

Explosive burnings occurring in C-O WDs produce the following nucleosynthesis products:

- *C-detonation*: mostly ^{56}Ni,

- *C-deflagration*: a large fraction of ^{56}Ni but also intermediate mass elements such as Ca, Ar and S plus O, Si, Ne, Mg and C,

- *He-detonation*: mostly ^{56}Ni.

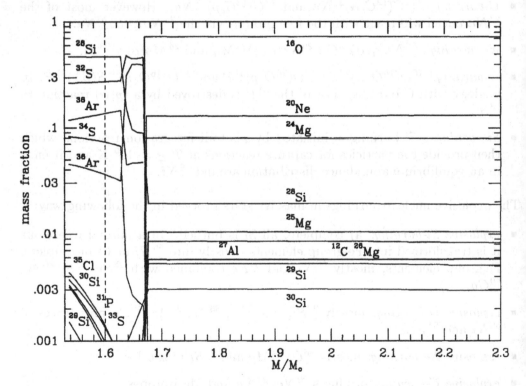

Figure 2.5. Mass fractions of major nuclei, resulting from hydrostatic burnings prior to core collapse for a star of initially $20M_\odot$. Displayed are the zones which will later be ejected by the propagating shock front. The separation between the remnant and the ejecta is indicated by a vertical dashed line. From Thielemann et al. 1990, Ap.J. Vol. 349, 222; reproduced by kind permission of F.K. Thielemann and the University of Chicago press (copy right 1990).

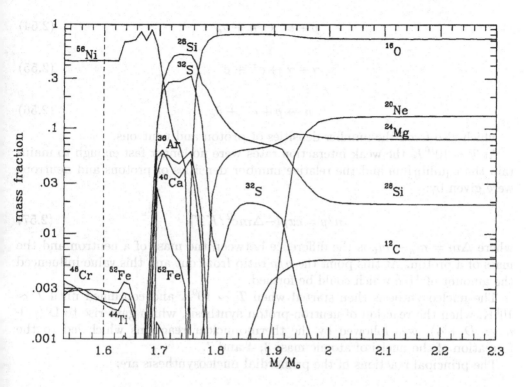

Figure 2.6. Mass fractions of a few major nuclei, resulting from explosive processing after the passage of a shock front for a star of initially $20M_\odot$. Matter outside $2M_\odot$ is essentially unaltered. Inside 1.7 $M_\odot$ the Fe-group elements dominate. The dashed line indicates the position of the mass cut if only $0.07M_\odot$ of ^{56}Ni are ejected, as indicated by SN 1987A. From Thielemann et al. 1990, Ap.J. Vol. 349, 222; reproduced by kind permission of F.K. Thielemann and the University of Chicago press (copy right 1990).

2.2.7   THE PRODUCTION OF LIGHT ELEMENTS

The light elements H, D, ^3He, ^4He and ^7Li (nuclei with A=5 and A=8 are unstable) are produced during the Big Bang although He and ^7Li are also produced in stars. The interaction between cosmic rays and CNO interstellar nuclei, called *spallation*, is responsible for the formation of the nuclei of ^6Li, ^9Be, ^{10}B and ^{11}B.

We recall here briefly the main processes occurring during the first 100 seconds of the life of the universe.

At temperatures $T > 10^{12}$ K statistic equilibrium existed among the existing particles which were p, n, e^+, ν_e, $\tilde{\nu}_e$ and γ. The equilibrium was governed by the following reactions:

$$n + e^+ \leftrightarrow p + \tilde{\nu}_e \tag{2.53}$$

$$p + e^- \leftrightarrow n + \nu_e \qquad (2.54)$$

$$\gamma + \gamma \leftrightarrow e^+ + e^- \qquad (2.55)$$

$$n \rightarrow p + e^- + \tilde{\nu}_e \qquad (2.56)$$

which also fixed the number densities of proton and neutrons.

At $T \sim 10^{10} K$ the weak interaction rates were no longer fast enough to maintain the equilibrium and the relative number densities of protons and neutrons were given by:

$$n/p = exp(-\Delta mc^2/KT) \qquad (2.57)$$

where $\Delta m = m_n - m_p$ is the difference between the mass of a neutron and the mass of a proton. At this point the n/p ratio froze out and this value influenced the amount of ^4He which could be formed.

The nucleosynthesis then started when $T \sim 10^9$K and continued until $T \sim 10^8$K, when the reaction of neutron-proton synthesis, which gave rise to D ($p + n \rightarrow D + \gamma$), was followed by the thermonuclear reactions which led to the formation of the nuclei of atomic mass 3, 4 and 7.

The principal reactions of the primordial nucleosynthesis are:

$$D(n,\gamma)^3H; \quad D(D,p)^3H \qquad (2.58)$$

$$D(p,\gamma)^3He; \quad D(D,n)^3He \qquad (2.59)$$

$$^3He(n,p)^3H; \quad ^3H \rightarrow^3 He + e^- + \tilde{\nu}_e \qquad (2.60)$$

$$^3H(D,n)^4He; \quad ^3He(n,\gamma)^4He \qquad (2.61)$$

$$^3He(D,p)^4He; \quad ^3He(^3He,2p)^4He \qquad (2.62)$$

The dominant product of the Big Bang nucleosynthesis was ^4He and its abundance depended on the n/p ratio and was given by:

$$Y_p = \frac{2(n/p)}{[1 + (n/p)]} \qquad (2.63)$$

Finally, the nuclei of ^7Li and ^7Be were produced through the reactions:

$$^4He(^3H,\gamma)^7Li \qquad (2.64)$$

$$^4He(^3He, \gamma)^7Be. \tag{2.65}$$

The primordial nucleosynthesis depends basically upon two parameters: i) the baryonic density of the Universe and ii) the number of neutrino families, N_ν. The baryonic density governs the rate of the thermonuclear reactions and in particular the abundance of D is the most sensitive to the baryonic density. This is due to the fact that the baryon to photon ratio, η, controls the onset of nucleosynthesis through the deuterium formation. On the other hand, the number of neutrino families influences primarily the abundance of ^4He. The reason for this resides in the fact that an increase in the number of neutrino families leads to a larger energy density which accelerates the expansion of the universe with the consequence of having an earlier occurrence of the freeze-out of the weak interactions as well as of the n/p ratio. The freeze out of the n/p ratio, in fact, depends on the competition between the weak interaction rates and the expansion rate of the universe. A larger n/p ratio, in turn, leads to a larger final amount of ^4He, as shown in Eq. (2.63).

It is worth noting here that ^3He and ^4He are also produced in stars during the H-burning, as we have already seen, and that ^7Li can also, in principle, be produced in stars. As we have seen, this is possible through the reactions of the second branch of the p-p chain. In particular, ^7Li can be formed by the reactions:

$$^3He + {}^4He \rightarrow^7 Be + \gamma \tag{2.66}$$

giving rise to ^7Be which should be rapidly transported (by convection) into regions of lower temperatures where the ^7Li is formed through the decay:

$$^7Be + e^- \rightarrow^7 Li + \nu_e + \gamma \tag{2.67}$$

In particular, ^7Li should be brought to temperatures low enough ($T < 2.8 \cdot 10^6$K) to avoid the fusion of ^7Li with H which gives rise to two α particles. However, this particular situation, originally proposed by Cameron and Fowler (1971), and known as the *Be transport mechanism*, is not easy to find in nature although it may work in particular stellar evolutionary phases for example during thermal pulses in AGB stars, which are indeed considered as possible sites for ^7Li production. Moreover, Li-rich AGB stars and RG stars are observed, thus confirming this hypothesis. Another possible way of producing ^7Li is during explosive H-burning occurring, for example, during a nova outburst and explosion of very massive stars. Nucleosynthesis calculations on the outcome of nova outbursts have been made by Starffield et al. (1978), Boffin et al. (1993) and Josè and Hernanz (1998) while the ^7Li production from massive stars is due to Woosley et al. (1990). In particular, the ^7Li in supernovae should arise from the ν-process acting in supernovae of type II. The ν-nucleosynthesis occurs because high energy μ and τ neutrinos excite heavy elements and even helium to particle unbound levels. The evaporation of a single neutron or proton and the back reactions of these nucleons on other species present significantly alters the nucleosynthesis. In particular, ^7Li is produced by the reactions:

$$^4He(\nu,\nu'n)^3He(\alpha,\gamma)^7Be(e^-,\nu_e)^7Li \qquad (2.68)$$

occurring during the explosion of type II supernovae. The most important regions for ν-nucleosynthesis are the helium, carbon and neon shells where significant amounts of ^7Li, ^{11}B and ^{19}F are synthetized.

We conclude by recalling that the primordial abundances of D, He and ^7Li are of fundamental importance to impose constraints on the density parameter of the Universe, in particular the baryon density parameter Ω_B through the parameter η (baryon/photon ratio) defined above. In particular, the most recent estimates of η_{10} (in units of 10^{-10}) give (Olive, 1999):

$$1.55 < \eta_{10} < 4.45 \qquad (2.69)$$

which translates into:

$$0.006 < \Omega_B h^2 < 0.016 \qquad (2.70)$$

with $h = H_o/100$. These values of Ω_B are quite low and would suggest that most of the dark matter in the universe is non-baryonic.

2.2.8 CONTRIBUTIONS OF SINGLE STARS TO CHEMICAL ENRICHMENT

We can now put together what is known about stellar evolution and stellar nucleosynthesis to define the different contributions to the chemical enrichment of the ISM from stars of different masses, in particular:

- $M \leq 0.08\ M_\odot$, if they exist, will never ignite H and their luminosity is simply given by the gravitational energy liberated in their slow contraction. They are called brown dwarfs. They do not contribute to galactic enrichment except for locking up material. Their lifetimes, τ_m, correspond to several times the age of the universe.

- $0.08 \leq M/M_\odot < 0.5$, they ignite H in the center but their He-core becomes electron degenerate before getting hot enough to burn. They will never ignite He and will end up as He-white dwarfs. They also contribute only by locking up material, their $\tau_m >> 15 \cdot 10^{10}$ years.

- $0.5 < M/M_\odot < M_{HeF}$, they ignite He in an electron degenerate core (He-flash) and end their lives as C-O white dwarfs. Only for $M \geq 1\ M_\odot$ these stars contribute to galactic enrichment in ^4He, ^{14}N and heavy s-process elements such as Ba and Sr. Their τ_m go from $>> 15 \cdot 10^{10}$ to $\sim 10^9$ years.

- $M_{HeF} < M < M_w$, they ignite He in a He-non-degenerate core but develop an electron degenerate C-O core. They end their lives as C-O white dwarfs and contribute to ^4He, ^{12}C, ^{13}C, ^{14}N, ^{17}O and s-process elements, which are produced during the shell He-burning.

- $M_w < M < M_{up}$, depending on whether $M_w \sim M_{up}$ (mass loss and/or overshooting) these stars should ignite C in a degenerate core when their mass reaches the Chandrasekhar limit with a subsequent thermonuclear explosion. These SNe, if they exist, should be different either from type II SNe or type Ia SNe and it was suggested to call them SN I1/2 (Iben and Renzini, 1983). However, empirical estimates give $M_w \sim 8~M_\odot$ (e.g. Weidemann 1987) whereas theoretical estimates of M_{up} from models with mass loss and over-shooting give $M_{up} \sim 6.6~M_\odot$ (Maeder and Meynet, 1989) and $M_{up} = 4-5M_\odot$ (Marigo et al. 1996) against the classic value of $M_{up} = 8-9~M_\odot$. This means that type I1/2 SNe are very unlikely to occur and stars in this mass range end their lives as white dwarfs. The lifetimes of stars in the mass range, $M_{HeF} - M_{up}$, go from $\tau_m \sim 10^9$ to several 10^7 years. These stars eject their nuclearly processed material into the ISM during the AGB phase by means of quiescent mass loss and during the PN phase. Episodes of dredge-up in conjunction with thermal pulses bring to the surface freshly synthesized material (H and He-burning products). Envelope burning will then transform this fresh material, e.g. ^{12}C, into ^{14}N and ^{13}C. In this case, these two elements have a *primary* origin (Renzini and Voli, 1981). Stars in this mass range, if in binary systems, can give rise to type Ia SNe and be responsible for substantial iron production (see next section).

- Stars in the mass range $M_{up} < M/M_\odot \le 10 - 12$, contribute to galactic enrichment in heavy elements such as ^{14}N and ^{12}C and perhaps traces of oxygen. They should explode as type II SNe (e-capture SNe) and leave a NeOMg white dwarf as a remnant. Their lifetimes are of the order of several times 10^7 years.

- $10 - 12 < M/M_\odot < M_{SNII}$, they are responsible for producing the bulk of heavy elements such as ^{16}O, ^{20}Ne, ^{24}Mg, ^{28}Si, ^{32}S, ^{40}Ca and r-process elements. They end their lives as type II SNe since, when they die, they are still in possession of their H-rich envelopes.

- $M_{SNII} \le M/M_\odot \le 100$, these stars are Wolf-Rayet stars, characterized by a strong mass loss. They eventually explode but lacking their H-envelope their light curves and spectra are different from those of type II SNe. They are possibly the progenitors of type Ib SNe although recently other candidates such as massive stars ($M > 10\ M_\odot$) in close binary systems have been proposed. Wolf-Rayet stars contribute to ^4He, ^{12}C, ^{22}Ne, ^{14}N, and perhaps ^{18}O through stellar winds as well as to the heavier elements through the SN explosion. Their lifetimes are $\tau_m \sim 10^6$ years.

- $M > 100 M_\odot$, these stars either explode as *pair-creation* SNe or implode into black holes. In the first case they mostly contribute to the ^{16}O enrichment. Models with overshooting and mass loss indicate for pair-creation SNe a range of initial masses of 100-200 $M_\odot$. Their lifetimes are $\tau_m \leq 10^6$ years.

- $M > 200 M_\odot$ They produce mostly ^4He and traces of ^{15}N and ^7Li (Woosley et al. 1984) when they suffer total disruption by explosive H-burning.

2.3   THE EVOLUTION OF BINARY SYSTEMS

We do not intend to describe the evolution of all binary systems and for this we address the reader to more specific articles and books (e. g. Iben and Tutukov 1984; Vanbeveren et al. 1994). Here we recall a few fundamental facts concerning the evolution of particular binary systems which can be related to the formation of SNe and novae and are relevant for the chemical evolution of galaxies.

2.3.1   SUPERNOVAE I

Although there is still a large uncertainty in the existing models, it is believed that binary systems can give rise to supernova explosions producing a relevant amount of heavy elements, in particular to type I SNe (Ia, Ib and Ic) and to nova outbursts which can produce non negligible amounts of ^7Li and perhaps ^{15}N, ^{13}C and even some Ne, Na, Mg, Al and Si.

We discuss first type Ia supernovae, which are thought to originate from white dwarfs in binary systems exploding by C-deflagration when they achieve the Chandrasekhar mass as a result of accretion or merger with the companion star. In fact, C-deflagration in C-O WDs whose mass exceeds the Chandrasekhar limit can reproduce many of the observed features in typical type Ia SNe, since it produces the right amount of $^{56}Ni \rightarrow^{56} Co \rightarrow^{56} Fe$ able to power their light curve ($\sim 0.5-0.6 M_\odot$) and allows for the formation of intermediate mass elements (from C to Si) which are observed in their spectra.

Up to now, two main models of binary systems leading to this scenario have been proposed:

- A DEGENERATE C-O WD PLUS A RED GIANT OR MAIN-SEQUENCE COMPANION which transfers H-rich material on the WD, the so-called "single degenerate model". A C-deflagration occurs when the WD (the primary mass of the binary system, namely the primordially more massive star) has accreted enough mass from the companion (the secondary star, namely the primordially less massive) to reach the Chandrasekhar mass limit. The most restrictive aspect of this scenario is that the accretion rate should be within a very narrow range of values in order to prevent the occurrence of H and He shell flashes (typical of cataclysmic variables) which prevent the WD from retaining the accreted mass and therefore reaching the Chandrasekhar mass. The clock for the explosion in this scenario is given by the lifetime of the secondary star which is in the range $0.8\text{-}M_{up}(M_\odot)$ and therefore it varies from the age of the universe to several 10^7 years. The expected realization frequency of these systems has been computed by means of population synthesis which follows the evolution of all the binary systems in the Galaxy. Recently Yungelson and Livio (1998) found this frequency to be only $\sim 10\%$ of the observed type Ia SN rate in the Galaxy (~ 0.003 SN yr^{-1}, Cappellaro et al. 1997). Some effort has been made recently to enlarge the range of possibilities for the single-degenerate scenario for type Ia SNe. Hachisu et al. (1996;1999) have proposed a particular single degenerate model where the secondary star can be either a MS (slightly evolved) star or a red giant star and have shown that if the accretion rate exceeds a certain limit, the WD blows a strong wind which stabilizes the mass transfer for a wider range of mass accretion rates than previously found and allows the WD to burn hydrogen steadily and to increase its mass, thus reaching the Chandrasekhar mass and exploding. The suggested progenitor masses for the primary stars are in the range $3 - 8 M_\odot$, while the progenitors of the secondary are $0.9 - 1.5 M_\odot$, in the case in which the secondary is a red giant, and $1.8 - 2.6 M_\odot$, in the case in which the secondary is a slightly evolved MS star. They showed that the expected frequency of SN Ia originating from such systems is in very good agreement with the observational galactic frequency. They also suggested a metallicity effect in the production of SN Ia, in the sense that only objects with [Fe/H] $\geq$ -1.0 dex may give rise to these SNe, the reason residing in the fact that, if the accreting material has abundances lower than that, the wind blown by the WD is too weak to stabilize the mass transfer and allow the steady H-burning which increases the WD mass above M_{Ch}. Also in the case of the Hachisu et al. model the clock for the explosion is given by the lifetime of the secondary star. This particular model would imply a longer timescale for the first occurrence of type Ia SNe with consequences on the chemical evolution of the Galaxy which still need to be tested.

- TWO DEGENERATE C-O WHITE DWARFS merging after gravitational wave emission, the so called "double degenerate model" (Iben and Tutukov 1984). The explosion mechanism is again C-deflagration. The clock for the explosion is given by the lifetime of the secondary star plus the gravitational time delay which can vary from 10^8 years to several Hubble times. The masses of the components of such a binary system are assumed to be almost the same and in the mass range $5 - M_{up}(M_\odot)$. Before becoming WDs the stars evolve through two common envelope phases at the end of which all the material in the common envelope is lost. After these two-common envelope phases, the system is made of two WDs which will gradually merge, as a result of the loss of angular momentum due to gravitational wave emission, and eventually give rise to a SN Ia event. In particular, Figure 2.7 demonstrates the evolution of such double degenerate system to a supernova Ia as indicated by the model of Iben and Tutukov (1984). In this model, the system originally consists of two Main-Sequence stars of comparable mass (5-9 $M_\odot$), semimajor axes between 70 and 1500 $R_\odot$ and orbital periods in the range 1 month to 6 years. The angular momentum carried away by the emission of gravitational waves forces the two components to be closer together over a timescale ranging from 10^5 to 10^{10} years and when the separation is reduced to $\sim 0.01 R_\odot$ the less massive of the two WDs is transformed into a heavy disk or rapidly rotating envelope around the more massive WD. When accretion of matter from the disk raises the mass of the WD above $\sim 1.44 M_\odot$ then the SN Ia occurs. The clock for the explosion in this scenario is given by the lifetime of the secondary star plus the gravitational time delay (see eq. 5.81), and it varies from several 10^7 years up to 10^{10} years and more. The most serious uncertainties in this scenario are: the fact that the merger of two WDs may lead to an accretion-induced collapse which can in turn trigger the formation of a neutron star or a white dwarf, if the merger is accompanied by mass loss, and the fact that no close binary WD system with a total mass higher than M_{Ch} has ever been found.

- SUB-CHANDRASEKHAR WD IN BINARY SYSTEMS

The fact that a sub-Chandrasekhar mass can still give rise to a SN explosion can find support in the observational findings of some sub-luminous type Ia SNe (see chapter 1). In the proposed scenario, a C-O WD accumulates an He-layer at low rates from a binary companion and ignites He off-center (at the bottom of the accreted layer) which results in a detonation before reaching the Chandrasekhar mass (see Livio 2000 for references).

In spite of the various possibilities discussed above and the uncertainties still existing in the evolution of binary systems and explosive nucleosynthesis, there is a general consensus on the basic model for the progenitors of typical type Ia SNe: *the thermonuclear disruption of C-O white dwarfs accreting H-rich material from a companion when they reach the Chandrasekhar limit and ignite C through a deflagration in their centers.*

Galactic chemical evolution models (see chapters 5 and 6) can impose constraints on the progenitors of SNe, in particular by means of the study of the relative abundance ratios with respect to the iron abundance.

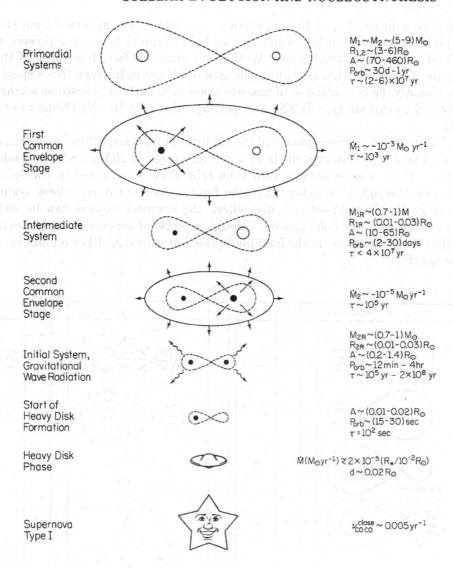

Figure 2.7. The main stages of the evolution of a system of two WDs merging after loss of angular momentum due to gravitational wave emission. The system evolves through two common envelopes phases before the two components become WDs. M_1 and M_2 are the primary and secondary mass, respectively. R_1 and R_2 are the radii of the primary and secondary star, respectively. R_{1R} and R_{2R} are the radii of the primary and secondary star after the common envelopes phases, respectively. The masses M_{1R} and M_{2R} are the mass after the two common envelope phases of the primary and secondary star, respectively. A is the separation of the system and τ is the timescale for the various phases. From Iben and Tutukov 1984, Ap.J.Suppl. Vol. 54, 335; reproduced by kind permission of I. Iben and the University of Chicago Press (copy right 1984).

86

The progenitors of type Ib,c supernovae are even more uncertain than those of type Ia supernovae. It has been suggested that type Ib SNe can represent the death of single Wolf-Rayet stars. Wolf-Rayet stars, in fact, lose most of their envelope of H and He thus explaining the lack of H lines in the type Ib SN spectra. More recently, Fe core collapse of massive stars in close binary systems seems to be favored to explain type Ic SNe and perhaps also type Ib SNe (Nomoto et al. 1995).

In Figure 2.8 the various phases of the evolution of a massive binary system are shown. The initial system is made of a primary mass of $20M_\odot$ and a secondary of $8M_\odot$. These masses change with time (the time is indicated in the second column on the right), as indicated in the figure. As one can see, these systems end up as double compact systems, where the compact objects can be either neutron stars or black holes passing through a phase of supernova explosion and of active X-ray emission. In the final phase the two stars can either remain bound or fly apart.

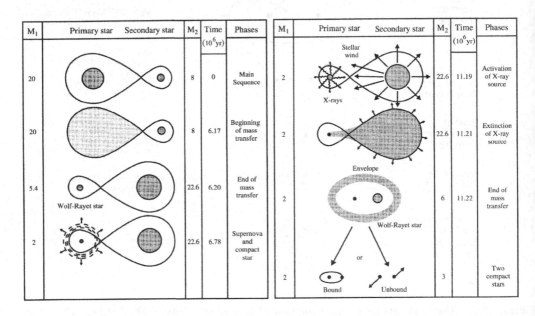

Figure 2.8. The evolution of a massive binary systems. The primary star is originally a $20M_\odot$ star and the secondary is a $8M_\odot$ star. From Luminet 1993, in "Black Holes"; reproduced by kind permission of J.P. Luminet and Cambridge University Press (copy right 1993).

2.3.2 NOVAE

The same binary systems which can give rise to SN Ia (single degenerate model) can produce nova events, depending on the rate of accretion on the white dwarf. In particular, the nova explosion originates from a system made of a C-O or ONeMg WD accreting H-rich material from a MS companion which fills its Roche lobe in a close binary system. The nova mechanism involves a surface thermonuclear explosion. The gas which is accreted by the white dwarf from the younger companion is compressed and heated by release of gravitational energy until H ignites explosively. Following this event, called "nova outburst" some material is lost. The occurrence of a nova or a supernova event in these systems depends on the rate of accretion from the companion: in fact, if the accretion rate is too fast, a nova event is produced with consequent loss of material thus preventing the white dwarf from reaching M_{Ch}, as discussed in the previous sections. A binary system of this kind can suffer $\sim 10^4$ outbursts during its lifetime and the first outburst does not occur before a timescale of the order of ~ 1 to several Gyr from the formation of the white dwarf.

2.4   STELLAR YIELDS

The word *yield* is probably misleading since it was originally introduced to indicate the ratio between the amount of chemical elements newly created and ejected by a stellar generation (all stars from $1M_\odot$ to a maximum mass) relative to the material which remains locked up in low mass never evolving stars and remnants (Tinsley, 1980). This definition implies the hypothesis of instantaneous recycling approximation (I.R.A.), namely that all the stars with mass $m \geq 1M_\odot$ have a negligible lifetime and that all the stars with $m < 1M_\odot$ live forever. While the latter statement is certainly true the former is not, but this approximation is necessary to solve analytically the equations representing the evolution of gas and metals (see chapter 5). Nowadays the term *stellar yield* is mainly used to indicate the masses of fresh elements produced and ejected by a star of initial mass m and metallicity Z, namely $(M_{ej})_i$ where i indicates a specific chemical element.

2.4.1   YIELDS FROM MASSIVE STARS

Massive stars are responsible for the creation of the bulk of the heavy elements (with the exception of the iron-peak and s-process ones) and in particular of oxygen which is the predominant element in the global metallicity Z. Since the explosion of supernova 1987A, a great deal of theoretical work has been devoted to the nucleosynthesis in massive stars $(m > 10M_\odot)$, neglecting perhaps the nucleosynthesis in low and intermediate mass stars $(1\text{-}8M_\odot)$. In the past years the stellar yields were usually calculated only for solar metallicity but more recently metallicity-dependent yields have become available. The most exhaustive study of metallicity-dependent nucleosynthesis in massive stars, at the time of writing, is from Woosley and Weaver (1995) (WW95). They calculated the yields of isotopes lighter than A=66 (zinc) for a grid of stellar masses and metallicities including stars of 11, 12, 13, 15, 18, 19, 20, 22, 25, 30, 35 and 40 $M_\odot$ and metallicities Z=0, 10^{-4}, 0.01, 0.1 and 1 times solar. The nucleosynthesis occurring during the supernova (II) explosion was also calculated in detail. They suggested that stars with massese larger than $30M_\odot$ experience considerable reimplosion of heavy elements following the start of the shock due to the explosion. This leads to a dramatic reduction of the yields of heavy elements and may leave a black hole remnant of up to 10 $M_\odot$ and more. They also suggested that the formation of black holes should be favored in stars of low metallicity because of their more compact structure and reduced mass loss. These new yields differ from the previous ones mostly because they include detailed calculations of explosive nucleosynthesis. Arnett (1995) has critically reviewed the new projects concerning explosive nucleosynthesis calculations. In particular, he compared the results of WW95, Thielemann et al. (1996, TNH95) and Arnett and concluded that the general agreement is surprisingly good, given the uncertainties still present in stellar calculations such as the rate of mass loss in the pre-SN phase, the treatment of convection, the rate of the $^{12}C(\alpha, \gamma)^{16}O$ reaction and the explosion mechanism. Generally the test for nucleosynthesis calculations is represented by their ability in reproducing the observed abundance ratios in Pop II stars, since these ratios should reflect the signature of the nucleosynthesis in massive stars, and Arnett again concluded that this comparison is also promising.

The importance of mass loss processes during stellar evolution in very massive stars $(M > 30 - 40M_\odot)$ has been investigated by several authors (e.g. Maeder 1992, M92); Langer and Henkel (1995) (LH95); Woosley et al. (1993, 1995). These stars become Wolf-Rayet and eventually explode possibly giving rise to peculiar type I supernovae (Ib, Ic). From the point of view of the chemical enrichment of the ISM they contribute to heavy elements but mostly to ^4He. In particular, as shown by M92, this should be true especially for stars of initial solar metallicity which would suffer greater loss of He, given the assumed dependence of the mass loss rate on the initial stellar metallicity.

In Figures 2.9 - 2.15 we show a comparison between yields from stars of different masses (from 10 to 40 $M_\odot$) and different metallicities (from 0.0 to solar) together with a comparison between yields from different authors for the solar metallicity. The comparison is made for some of the most important heavy elements (C, N, O, Mg, Si and Fe) and for He. For this last element the yields calculated by taking into account mass loss are available, and from the comparison with the other studies without mass loss one can see that the effect of mass loss is to increase the amount of He which is ejected into the ISM and therefore subtracted from further nuclear processing. As a consequence of that, the net effect of mass loss is to increase the He yields and to depress the Z yields. This fact has important consequences on the He enrichment relative to metals during the galactic lifetime ($\frac{\Delta Y}{\Delta Z}$). From the figures one notices that the yields from different sources differ by no more than a factor of two which is remarkable, as already noticed. On the other hand, the largest differences are between the yields for $Z = 0$ and those for $Z \neq 0$ in WW95 calculations. However, it should be said that from the point of view of galactic evolution this is not very important since very few stars of zero metallicity must have existed, giving the fact that a few SN explosions must have raised the Z content to a value different from zero in a few million years.

Concerning the yields relative to a solar chemical composition without mass loss from different authors (WW95 and TNH95), the main differences reside in the "criterion" adopted for convection, the nuclear reaction rates and the "mass-cut", namely the mass which defines the border between the remnant and the ejecta. The main differences among these two sets of yields is in the mass of Fe: in the case of the yields of WW95 the iron increases with stellar mass whereas in the case of TNH95 all massive stars eject $\sim 0.07 M_\odot$ of Fe. If we take all of the uncertainties into account we can say that the stellar yields are uncertain by a factor of ~ 2 for oxygen and by larger factors for the other isotopes. Particularly uncertain is the yield of Fe and perhaps we should be guided by the observations of Fe produced by core-collapse SNe in order to constrain the theoretical Fe yields of massive stars (see Figure 1.26).

2.4.2 YIELDS FROM LOW AND INTERMEDIATE MASS STARS

These stars, as is well known, contribute through quiescent mass loss and the PN phase to the enrichment in ^4He, ^{12}C, ^{13}C, ^{14}N, ^{17}O and heavy s-process elements. As noted before, in the years following the work of Renzini and Voli (1981) most of the attention was devoted to the massive stars and only recently new studies of nucleosynthesis in low and intermediate mass stars have appeared (van den Hoek and Groenewegen, 1997; Marigo et al. 1996).

The major source of uncertainty affecting the nucleosynthesis in low and intermediate mass stars is the mixing-length parameter α_{ML}, which determines the depth of the convective stellar envelope and the extent of the hot-bottom burning. Particularly important but at the same time uncertain are the predictions for the production of N, the bulk of which should originate in low and intermediate mass stars. Nitrogen is important since it is mainly produced as a *secondary* element during H- burning (CNO cycle), and this fact implies that its abundance should increase in a way different from that of other elements produced by the same stars, such as carbon, but as *primary* elements. The different evolutionary rate of N relative to other elements, in fact, can be used to infer the age and the nature of galaxies as discussed in the following sections.

Renzini and Voli (1981) pointed out that part of N in intermediate mass stars can have a primary origin since it derives from the CNO cycle using freshly made C and O, as a result of the third dredge-up acting during the AGB phase. The size of this primary fraction of N is quite uncertain since it depends critically

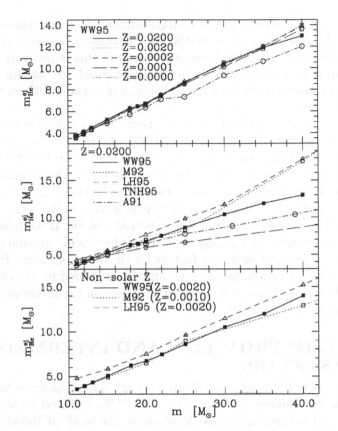

Figure 2.9. Yields of ^4He as functions of the initial stellar mass for different initial metallicities as computed by different authors (WW95, LH95, TNH95, M92). A91 stands for Arnett (1991). Courtesy of B.K. Gibson.

upon the treatment of convection. A study of Blöcker and Schönberner (1991) suggested that the third dredge-up could possibly not even occur. In such a case the amount of primary N produced in intermediate mass stars should be strongly reduced. The results of the new studies do not change substantially the situation relative to the previous yields, although the amounts of produced and ejected ^{14}N and ^{13}C are lower.

In principle, there could be some primary production of N in massive stars, as often advocated for several astrophysical situations (Matteucci 1986), and in principle it could happen as in AGB stars (dredge-up plus hot-bottom-burning) but again the uncertainty resides in the treatment of convection. Recently, Heger (1998) and Maeder (1999) have shown that rotationally induced mixing can induce primary production of N in massive stars.

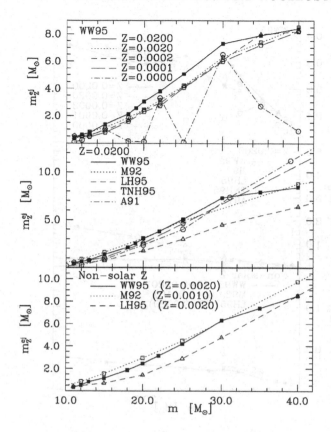

Figure 2.10. Yields of the global metallicity Z for different initial metallicities as computed by different authors, as in Figure 2.9. Courtesy of B.K. Gibson.

2.4.3   YIELDS FROM TYPE I SUPERNOVAE

As we have already discussed, type Ia SNe are commonly believed to originate from exploding white dwarfs in binary systems, although the nature of their progenitors is still in question. As we have already discussed, the most popular model is the C-deflagration of a C-O white dwarf triggered by accretion of material from a companion. The companion can either be a red giant or another white dwarf .

Nomoto et al. (1984) predicted the yields for type Ia SNe deriving from the C-deflagration of a C-O white dwarf reaching the Chandrasekhar mass. These yields have been very popular since they seem to reproduce very well the observed abundances in type Ia SN spectra (Branch and Nomoto 1986; Mazzali et al. 1993), especially those predicted by their model W7 (see also Figure 2.4). Thielemann et al. (1993) and Iwamoto et al. (1999) presented an updated version of the Nomoto et al. yields. The differences between the old and new yields

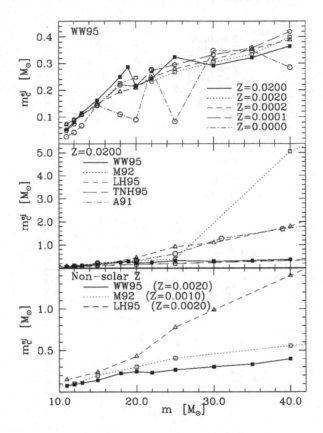

Figure 2.11. Yields of ^{12}C computed for different initial metallicities and by different authors. Courtesy of B.K. Gibson.

mainly concern Ne and Mg but do not affect appreciably model predictions. We note here that these SNe produce mostly iron ($\simeq 0.626 M_\odot$ per SN, as shown in Table 2.1). In Table 2.1 we show the masses of few major chemical species produced by a SN Ia, as predicted by model W7. As one can see, Fe is the most abundant species produced, whereas O, Ne and Mg are produced in negligible amounts if compared with the production by massive stars. In order to have a comparison with the massive star production we report in Table 2.1 also the average yields for type II SNe obtained by weighting the yields of Iwamoto et al. (1999) in the range 10-50$M_\odot$ on the Salpeter (1955) initial mass function (see chapter 3). This table clearly shows that the bulk of the typical α-elements with the exception of Si, S and Ca are originating from massive stars.

SNe of type Ia, because of their progenitors, explode on a large range of timescales, ranging from $3 \cdot 10^7$ years to a Hubble time for the model with the single white dwarf and from several 10^8 years to several Hubble times for the

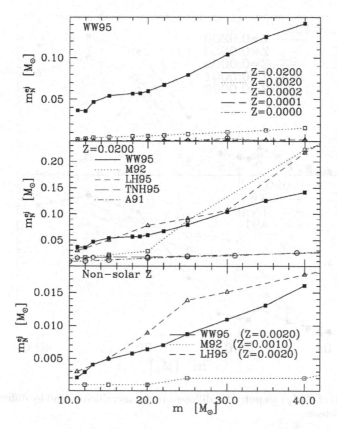

Figure 2.12. Yields of ^{14}N computed for different initial metallicities and by different authors. Courtesy of B.K. Gibson.

model with the double white dwarfs. Therefore, we are expecting that the iron produced in type Ia SNe will be restored into the ISM with a significant delay relative to other elements such as oxygen which are produced and ejected on very short timescales by massive stars. This delay is even larger if we consider possible metallicity effects preventing a stable accretion from the companion for metallicities below a certain threshold, as recently suggested by Hachisu et al. (1999).

Concerning type Ib SNe the situation of the progenitors is even less clear, although there is now a general consensus in believing that they could be the result of the explosion of Wolf-Rayet stars either single or in binary systems. From the nucleosynthetic point of view the type Ib SNe should produce less iron than type Ia SNe, as inferred from their light curves which show a maximum luminosity lower by a factor from 1.5 to 2.0 magnitudes than type Ia SNe, but detailed nucleosynthesis calculations for these SNe are not available.

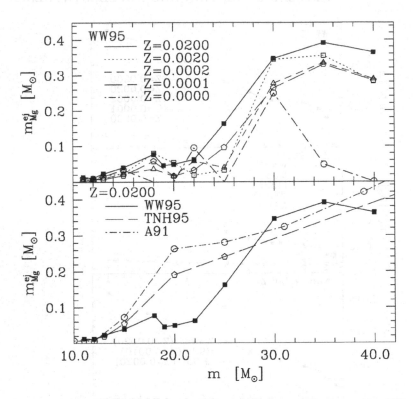

Figure 2.13. Yields of ^{24}Mg computed for different initial metallicities and by different authors. Courtesy of B.K. Gibson.

2.4.4 YIELDS FROM NOVAE

Theoretical calculations have shown that novae can produce several nucleids such as ^7Li, ^{13}C, ^{15}N and ^{17}O as well as some radioactive isotopes (^{22}Na, ^{26}Al) (see Gehrz et al. 1998, for a review on the subject), all products of explosive H-burning. Novae can in principle be important ^7Li producers in the Galaxy, as we will see in chapter 6.

2.4.5 SUMMARY OF ELEMENT PRODUCTION IN STARS

After having discussed the most important ingredients in the stellar evolution models and the sites of production of the different chemical species and before summarizing what we said in the previous sections we note the general uncertainties still present in nucleosynthesis and stellar evolution. They are: 1) the rate of the nuclear reaction $^{12}C(\alpha, \gamma)^{16}O$, 2) the treatment of convection, 3) the nature of the explosion mechanism in massive stars (prompt or delayed), 4) the range

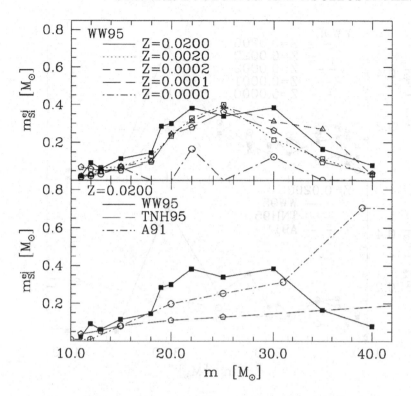

Figure 2.14. Yields of ^{28}Si computed for different initial metallicities and by different authors. Courtesy of B.K. Gibson.

of stellar masses which do indeed explode (which is the limiting mass M_{bh} for forming a black hole?), 5) the velocity of the deflagration front in SN Ia models.

Where and how the elements are produced?

- ^1H is created during the Big Bang and is only destroyed during galactic evolution,

- D, ^3He, ^4He, ^7Li are produced during the Big Bang, although He and ^7Li are also partly produced in stars,

- D is only destroyed during galactic evolution,

- ^3He is mainly destroyed but also produced, although in small quantity, in stars in the mass range $1-2M_\odot$,

- ^4He is produced in stars with masses between 1 and 100 $M_\odot$ and its production strongly depends on the amount of mass loss in stars,

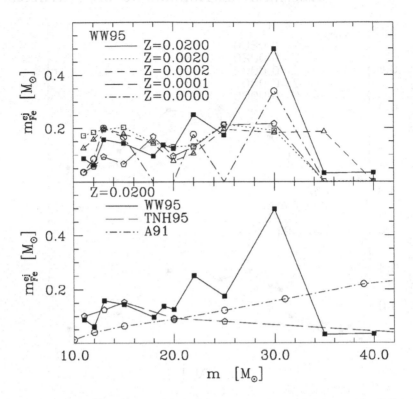

Figure 2.15. Yields of ^{56}Fe computed for different initial metallicities and by different authors. Courtesy of B.K. Gibson.

- ^{7}Li is probably produced in AGB stars (5-$8M_{\odot}$) and perhaps in novae and supernovae of type II. A fraction of ^{7}Li is produced by spallation but the exact amount is still uncertain.

- ^{6}Li, ^{9}Be, ^{10}B, ^{11}B are produced by spallation of cosmic rays on atoms of the ISM,

- ^{12}C is produced during the quiescent He-burning in high and intermediate mass stars,

- ^{13}C is produced during the quiescent and explosive H-burning (cold CNO cycle $T \leq 5 \cdot 10^{7}$ K, warm CNO cycle $T \sim 1.2 \cdot 10^{8}$ K) both in single and binary intermediate mass stars (novae). ^{13}C can be a partially primary element as a result of the third dredge-up and hot-bottom burning during the AGB phase of intermediate mass stars,

Table 2.1.   Masses ($M_\odot$) of few major species predicted by the model W7 and compared with the yields from massive stars averaged on the stellar mass according to the Salpeter IMF (Iwamoto et al. 1999).

Chemical element	W7	SNII
^{12}C	0.048	0.0794
^{16}O	0.143	1.8000
^{20}Ne	0.002	0.2120
^{24}Mg	0.009	0.0880
^{28}Si	0.154	0.1050
^{32}S	0.085	0.0380
^{36}Ar	0.014	0.0066
^{40}Ca	0.012	0.0057
^{48}Ti	2.05(-4)	1.16(-4)
^{56}Fe	0.626	0.0844

- ^{14}N is produced during the quiescent H-burning (cold CNO) in low and intermediate mass stars. ^{14}N can also be a primary element like ^{13}C,

- ^{15}N is produced in explosive H-burning occurring in SNe and novae ($^{15}N/^{14}N$ in cold CNO is 100 times lower than $^{15}N/^{14}N$ in the solar system),

- ^{16}O is produced during the He-burning in massive stars,

- ^{17}O is produced during the cold CNO-burning in single low and intermediate mass stars ($M > 4\ M_\odot$) (also in explosive burning),

- ^{18}O is produced from destruction of ^{14}N by $^{14}N(\alpha, \gamma)^{18}F(\beta^+)^{18}O$ occurring in stellar regions suffering He-burning. Restored into the ISM from SN II. Produced also in quiescent and explosive H-burning.

- ^{20}Ne, ^{24}Mg are produced during the C-burning in massive stars and the C-deflagration in C-O white dwarfs (only a very small amount). Mg is produced also during the quiescent and explosive Ne-burning,

- ^{28}Si, ^{32}S are produced during the O-burning in massive stars (quiescent and explosive) and C-deflagration in C-O WDs,

- ^{40}Ca is produced during the explosive O-, Si-burnings in massive stars and C-deflagration in C-O WDs,

- ^{56}Fe is produced during the Si-burning (quiescent and explosive) in massive stars and C-deflagration in C-O WDs (substantial contribution),

- s-process elements are produced during the He-burning in massive stars $^{22}Ne(\alpha, n)^{25}Mg$ (A< 90) and He-shell flashes $^{13}C(\alpha, n)^{16}O$ (A> 90) in low mass stars,

- r-process elements are produced during the explosive He, C, O or Si burning in type II SNe or in the very rich neutron matter originating from neutron stars.

2.5 LOCAL ABUNDANCES AND NUCLEAR PROCESSES

The local abundance distribution (often referred to as "cosmic abundance distribution", see chapter 1) contains the imprint of the various processes of nucleosynthesis. In Figure 2.16, abundance data based on a combination of elemental and isotopic determinations in the Solar System with those from nearby stars and emission nebulae are shown.

In particular, the curve in Figure 2.16 shows a number of peaks and features which give clues to the origin of the different elements:

- *Hydrogen* is the most abundant element followed by helium. H is entirely formed during the Big Bang whereas a small fraction of He is produced by stars (the primordial He abundance is $Y_P \sim 0.24$).

- *Li, Be, B* are very scarce since they are destroyed in stars although some of them can also be created there. They are much more abundant in primary cosmic rays.

- *D*, although still more fragile than Li, Be and B, is largely more abundant. D is formed during the Big Bang and only destroyed in stars so that its observed abundance represents a lower limit to the primordial abundance.

- *Nuclei from Carbon to Calcium* (A=12 up to A=40) form a group whose major structure appears to be a downward slope, owing to the progressive onset of the various nuclear burnings, with the odd-even effect superimposed. The even A nucleids in this group can be attributed to the burning of helium to carbon and carbon to heavier species, primarily through addition of α-particles.

- *Iron-group elements* show a peak at A=56 stretching from $A \approx 45$ to $A \approx 67$, owing to the fact that ^{56}Ni, the double magic nucleus which later decays into ^{56}Fe, is produced in conditions of NSE. Moreover, in correspondance of

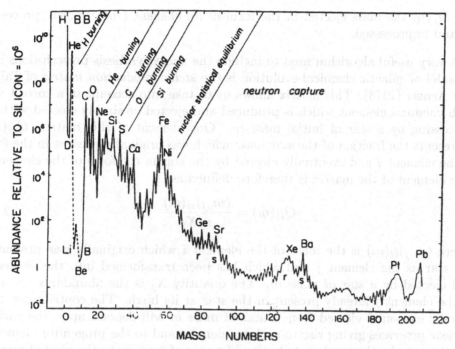

Figure 2.16. The local galactic abundance distribution of nuclear species, normalized to 10^6 ^{28}Si atoms. The main nuclear processes responsible for the production of elements are indicated. From Pagel 1997, "Nucleosynthesis and Chemical Evolution of Galaxies"; reproduced by kind permission of B.E.J. Pagel and the Cambridge University Press (copy right 1997).

^{56}Fe the binding energy per nucleon reaches a maximum, thus preventing the synthesis of heavier elements. When one uses this formalism

- *Neutron capture processes* can give rise to the so-called magic number peaks which follow the iron-peak. They correspond to closed shells with 50, 82 or 126 neutrons.

2.6   ALGORITHMS TO COMPUTE THE STELLAR YIELDS

The stellar yields depend upon a few fundamental parameters:

- M_α, the mass of the He-core (where H is turned into He)

- M_{CO}, the mass of the carbon-oxygen (C-O) core (where He is turned into heavier elements)

- M_R, the remnant mass

- $(M_{ej})_i$, the mass ejected in the form of an element i both newly processed and unprocessed.

A very useful algorithm used to include the nucleosynthesis prescriptions into a model of galactic chemical evolution is the *stellar production matrix* of Talbot and Arnett (1973). This matrix allows us to take into account in a correct way each chemical element which is produced and ejected or simply ejected without processing by a star of initial mass m. One element of this matrix, $Q_{ij}(m)$, represents the fraction of the star mass which was originally present in the form of the element j and eventually ejected by the star in the form of the element i. One element of the matrix is therefore defined as:

$$Q_{ij}(m) = \frac{(m_{ej})_{ij}(m)}{mX_j} \qquad (2.71)$$

where $(m_{ej})_{ij}(m)$ is the mass of the element i which originally was present in the star as the element j and which has been transformed into the element i and ejected by a star of mass m. The quantity X_j is the abundance by mass of the element j already present in the star at its birth. The contribution to a specific element i ejected from a star of mass m will depend upon the various nuclear processes giving rise to such an element and to the progenitor elements. In other words, the total contribution of a star of mass m to the ejected mass of the element i, both newly formed and already present, is given by:

$$(M_{ej})_i = \sum_{j=1,n} Q_{ij}(m)X_j m. \qquad (2.72)$$

For the sake of simplicity, the mass loss is regarded as taking place at the end of the stellar life. When one uses this formalism in a galactic chemical evolution model, the abundance X_j is the abundance of the ISM at the moment of the star birth. In general, the ij matrix element may depend upon the initial chemical composition, although most of the influence of composition upon $(M_{ej})_i$ is exhibited explicitly by the linear dependence upon X_j.

In order to define the stellar production matrix for a star of mass m we need to define several quantities:

- $d = \frac{M_R}{m}$ mass fraction eventually forming the stellar remnant

- $q_4 = \frac{M_\alpha}{m(X_H + X_D + X_{3He})}$ mass fraction involved in the nuclear H-burning $(H \rightarrow^4 He)$

- $q_3 = \frac{M_{3He}}{mX_{3He}}$ mass fraction where each original ^3He is transformed into ^4He or heavier species

- $q_c = \frac{M_{CO}}{m(X_H+X_D+X_{He})}$ mass fraction involved in the nuclear burning converting ^4He into ^{12}C, ^{16}O and heavier elements

- $w_2 = \frac{(M_{ej})_D}{mX_D}$ mass fraction within which the original D is tranformed into ^3He and *ejected*

- $w_3 = \frac{(M_{ej})_{3He}}{m(X_H+X_D)}$ mass fraction of newly formed and *ejected* ^3He

- $w_4 = \frac{M_{Li}}{mX_{Li}}$ mass fraction in which each original ^7Li (very likely all of it) is transformed into ^4He and *ejected*

- $w_7 = \frac{(M_{ej})_{Li}}{m(X_H+X_D+X_{He})}$ mass fraction of the newly produced and *ejected* ^7Li

- $w_c = q_c - d$ mass fraction which has undergone He-burning and has been *ejected*

- $w_{Ns} = \frac{(M_{ej})_N}{m(X_{12_C}+X_{16_O})}$ mass fraction where the original ^{12}C and ^{16}O have been transformed into ^{14}N during the CNO cycle and *ejected*

- $w_{13C_S} = \frac{(M_{ej})_{13_C}}{mX_C}$ mass fraction where the original ^{12}C has been transformed into ^{13}C during the CNO cycle and *ejected*

- χ_i is the mass fraction of w_c which is *ejected* in the form of the individual species i (i can be any element from ^{16}O up to ^{56}Fe). It is worth noting that when ^{14}N and ^{13}C have a primary origin they are included into w_c. The sum of all these fractions is equal to unity, by definition.

- $-w_{Cu_s}$ and w_{Zn_s} represent the mass fractions where the original Fe is transformed into Cu and Zn, respectively, and *ejected*. These isotopes are taken as an example of elements beyond the Fe-peak which are partly formed by means of s-processing, and therefore we treat them as partly secondary elements. However, treating the s-process elements as secondary elements is too simplistic, since their production depends also on the neutron-flux which in turn depends on metallicity. Therefore, it is preferable, when possible, to use detailed yield calculations for these elements instead of the production matrix.

The single elements of the production matrix (see Table 2.2) should be defined in the following way: in the columns we indicate the *progenitor* chemical species, whereas in the rows we indicate the *products*. In order to have the $(M_{ej})_i$ for a star of initial mass m we have to multiply each term in the same row by the abundance of its corresponding progenitor and then sum all of the contributions in a row. At this point we still have to multiply the obtained quantity by m since the elements of the matrix have been divided previously by m. Note that the elements along the diagonal of the matrix represent the species ejected by the star without having been nuclearly processed, in other words the pre-existing species. In order to know if the matrix has been written in a correct way one should add all the elements of a column and the result should always give $1 - d$, as expected. This rather complex way of proceeding has many advantages, the first is that by means of the production matrix one can take into account the secondary nature of a chemical species automatically. For example, nitrogen is the by-product of the H-burning in the CNO cycle and therefore depends on the initial abundances of C and O in the star. In the production matrix this is taken into account when one multiplies by the original abundances of C and O.

As an example, in Table 2.2 we show a matrix for 19 chemical elements (H, D, ^3He, ^4He, ^7Li, ^{12}C, ^{13}C, ^{14}N, O, nr (neutron-rich elements), Ne, Mg, Si, S, Ca, Fe, Cu, Zn, Ni). The principle on which the single elements of the matrix are constructed is the following: let us take for example the element $Q_{1,1}(m) = 1 - q_4 - w_7 - w_3$ which represents the amount of unprocessed H which is restored into the ISM, and is obtained by subtracting from the initial amount of H, namely unity, the fractions of H which have been processed to give rise to ^3He, ^4He and ^7Li. In the same way we define $Q_{2,2}(m) = 1 - q_3 - w_2 - w_7$ which is the amount of unprocessed D restored into the ISM, since deuterium is only destroyed in stars. Actually this quantity is likely to be equal to zero since all of the preexisting D is probably destroyed. It is worth noting in Table 2.2 that the *primary* elements, such as ^{12}C and the α-elements, for example, have terms different from zero only in correspondance of H, D, ^3He and ^4He as progenitors, since they do not need the presence of pristine metals to be synthesized. On the other hand, *secondary* elements such as ^{13}C and ^{14}N have non-zero matrix elements in correspondance of their progenitors (^{16}O and ^{12}C in the case of ^{14}N and ^{12}C in the case of ^{13}C). However, ^{14}N and ^{13}C can also have a primary origin if synthesized starting from the freshly made ^{12}C and O during the hot-bottom burning in the AGB phase of intermediate mass stars. In this case, the primary component of these elements will appear in the columns corresponding to H, He and D. The tenth row corresponds to the so-called neutron-rich elements (such as e.g. ^{18}O, ^{22}Ne). They are indicated by nr and have non-zero matrix elements only in correspondence of their progenitors (^{12}C, ^{13}C, ^{14}N and ^{16}O), since they are only secondary elements. A secondary component is present also for Cu and Zn since they are produced in part as s-process elements and therefore depend on the original abundance of Fe. The matrix of Table 2.2 can be used for single low, intermediate and massive stars, whereas for supernovae of type Ia, which originate from binary systems and do not leave any remnant one can use the matrix of Table 2.2 computed for the Chandrasekhar (or sub-Chandrasekhar) mass of a C-O white dwarf; in particular, one should assume $d = 0$ and that all the C-O mass (i.e. w_c) should be transformed into Fe and elements from C to Si according to the yields of type Ia SNe (see Table 2.1). The yields of all the other elements, namely the elements ejected during the normal evolution of the primary star (the secondary star can be ignored since it reaches only the red giant phase and most of the significant elements in low and intermediate mass stars are ejected long after this phase), should be instead calculated normally according to the initial mass of the primary. However, at the present time the use of the matrix can be avoided since detailed nucleosynthesis calculations including both processed and unprocessed material are becoming available (e.g. WW95).

2.7   THE INFLUENCE OF MASS LOSS AND OVERSHOOTING ON THE STELLAR YIELDS

The mass M_α (mass of the He-core) and M_{CO} (mass of the C-O core) are generally controlled by the physics adopted in the stellar evolutionary models such as mass loss (stellar winds) and overshooting from convective cores.

2.7.1   MASS LOSS

It has never been clear whether the evolution of massive stars with mass loss and no overshooting leads to smaller or larger M_α masses than in the conservative case (no mass loss, no overshooting). In fact, the first study of this type (Chiosi et al. 1978, 1979) found a smaller M_α as a consequence of mass loss. Later, Maeder (1981,1983) did not find any sensitive difference relative to conservative models. Bressan et al. (1981), including mass loss and overshooting in their stellar models found that, owing to the competing effects of mass loss and overshooting, M_α was again similar to that in the conservative models since the two effects compensate each other. Maeder and Meynet (1987,1989) recomputed models with mass loss and moderate overshooting and found that M_α is larger than in the conservative case. More recently, M92 has shown that mass loss depends on metallicity and large mass loss rates imply a lower production of metals and a larger production of He. In figures 2.17 and 2.18 we display the stellar yields under various assumptions about the mass loss in massive stars.

The effect of mass loss in low and intermediate mass stars is that the stellar yields increase with decreasing mass loss rate, since the lifetime of the AGB phase is longer and thus more thermal pulses are occurring, as shown by the more recent calculations.

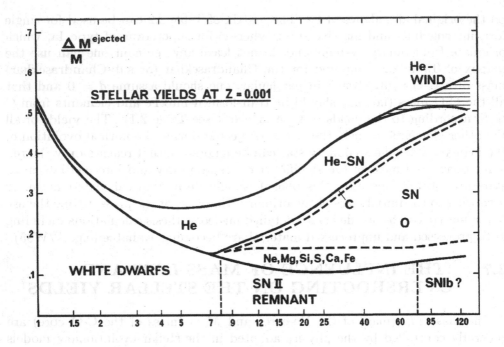

Figure 2.17. Mass fractions ejected as a function of the initial masses for metallicity Z=0.001. The quantities shown are the mass fractions $p_{im} = \frac{(M_{ej})_i}{m}$. The stellar wind contribution is indicated by hatched areas. Some indications about the composition of the ejecta are shown. The lower part represents the mass fraction in the remnants. From Maeder 1992, A & A Vol. 264, 105; reproduced by kind permission of Springer Verlag (copy right 1992).

2.7.2   CONVECTIVE OVERSHOOTING

The net effect of convective overshooting in stars of all masses is to increase M_α and M_{CO} with respect to the case with no overshooting. The increase of M_α and M_{CO} in massive stars leads to an increased production of heavy elements, whereas in intermediate mass stars $(M > M_{HeF})$ the situation is more complicated. Serrano (1986) pointed out that the effect of overshooting in intermediate mass stars results in a reduction of primary ^{14}N and ^{13}C as well as ^4He produced during the envelope burning. This is due to the larger core mass in the presence of overshooting for the same final mass which leads to a smaller mass in the envelope available for nuclear processing during the third dredge-up and envelope burning.

The effect of overshooting is also to decrease the value of M_{up} due to the increased core mass for the same initial mass, as we have already discussed. Nucleosynthesis calculations for intermediate mass stars $(0.7 - 4.0\ M_\odot)$ in the presence of strong overshooting and for two different stellar metallicities established a value of $M_{up} \sim 4 - 5\ M_\odot$ (Marigo et al., 1996). On the other hand, van

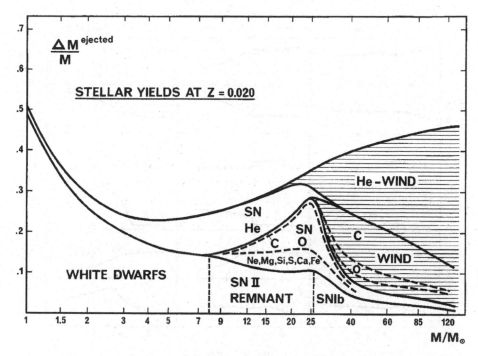

Figure 2.18. The same as Figure 2.17 but for Z=0.02. From Maeder 1992, A & A Vol. 264, 105; reproduced by kind permission of Springer Verlag (copy right 1992).

den Hoek and Groenewegen (1997) found a larger value of M_{up} and calculated new yields for low and intermediate mass stars in the range $0.8 - 8.0\ M_\odot$ for different stellar metallicities and compared them with the Renzini and Voli (1981) values (see Figure 2.19). The stellar models they adopted are from the Geneva group and contain moderate overshooting. The comparison with the old yields has shown that the new yields differ from the previous ones by a factor 2-3. In particular, for high mass AGB stars ($M > 3.5\ M_\odot$) the effect of the hot-bottom-burning on the ^{14}N yields is much lower than for the old yields. This implies a value of the mixing-length parameter, as defined in Renzini and Voli (1981), lower than 2 to be the most appropriate for massive AGB stars. An interesting aspect of these new calculations is that they show that the yields are strongly influenced by the initial stellar metallicity acting on the amount of dredged-up material, which is larger in initially low-Z AGB stars. This fact implies that C and O are higher at low Z whereas N increases *slightly* with Z.

2.8   GALACTIC YIELDS

By means of the stellar production matrix defined above and the nucleosynthetic calculations for stars of different masses, we can compute the *galactic*

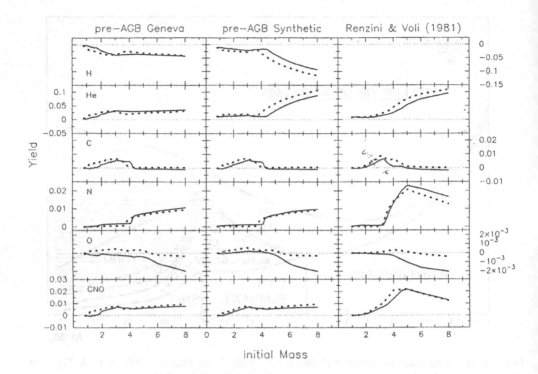

Figure 2.19. Stellar yields p_{im} for low and intermediate mass stars including hot-bottom-burning versus initial stellar mass of H, ^4He, ^{12}C, ^{14}N, ^{16}O and total CNO in the standard model of van den Hoek and Groenewegen (1997) for different treatments of the pre-AGB evolution (left and center panels) and initial metallicities Z=0.02 (solid line) and Z=0.004 (dotted line). The yields by Renzini and Voli (1981), their case $\alpha_{ML} = 2.0$, initial composition Z=0.02, Y=0.32, are shown for comparison. From van den Hoek and Groenewegen 1997, A & A Suppl. Vol. 123, 305; reproduced by kind permission of Springer Verlag (copy right 1997).

yields. As previously mentioned, the galactic yield of a chemical species i, as defined by Tinsley, is based on the I.R.A. and sudden mass loss approximation (S.M.L.A.) and is defined as: *the ratio between the total mass of the species i newly formed and ejected by all the stars larger than $1M_\odot$, assumed to die instantaneously, and the amount of mass locked up in low mass stars and remnants,* namely:

$$y_i = \frac{1}{(1-R)} \int_1^\infty m p_{im} \varphi(m) dm \qquad (2.73)$$

where $p_{im} = \frac{(M_{ej})_i}{m}$ with m being the initial stellar mass and $\varphi(m)$ is the so-called *initial mass function,* namely the distribution of stars according to their mass, which will be discussed in the next chapter. The quantity R is also defined under the I.R.A. and represents the total mass fraction, nuclearly processed and unprocessed, which is returned into the interstellar medium by a stellar generation (always all the stars from $1M_\odot$ up to infinity):

$$R = \int_1^\infty (m - M_R)\varphi(m)dm \tag{2.74}$$

This quantity is called fraction because is divided by $\int_0^\infty m\varphi(m)dm = 1$, which represents the normalization of the initial mass function (see section 3.2.3). The returned fraction R depends upon the initial mass function $\varphi(m)$ and the $m - M_R$ relation.

Tinsley assumed a $m - M_R$ relation such that:

$$M_R = 0.7M_\odot \quad m \le 4M_\odot \tag{2.75}$$

$$M_R = 1.4M_\odot \quad m > 4M_\odot \tag{2.76}$$

Later, Weidemann (1987) suggested the following semi-empirical relation:

$$M_R = 0.05m + 0.5 \quad 1 \le m/M_\odot < 6 \tag{2.77}$$

$$M_R = 0.144m \quad 6 \le m/M_\odot < M_{up} \tag{2.78}$$

In general, for stars more massive than M_{up} it is advisable to use $M_R = 1.4M_\odot$ although the most massive stars ($M > 30M_\odot$) can leave black holes as remnants with masses between 3 and 10 $M_\odot$. In Figure 2.20 is shown a comparison between empirical and theoretical initial-final mass relations for stars with initial mass in the range $1 - 8M_\odot$. As one can see from the figure, the more recent calculations reproduce the semi-empirical initial-final mass relations better than the predictions of Renzini and Voli (1981).

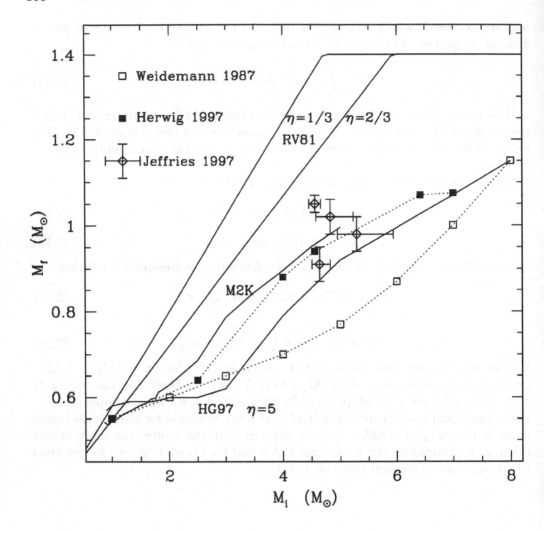

Figure 2.20. Initial-final mass relation for low- and intermediate-mass stars with solar metallicity. Semi-empirical calibrations for the solar neighbourhood are taken from Weidemann (1987), Herwig (1997), and Jeffries (1997). Solid lines refer to theoretical predictions for low- and intermediate mass-models with initial solar metallicity. In particular, the theoretical initial-final mass relations are by Renzini and Voli (1981, RV81), van den Hoek and Groenewegen (1997, HG97) and Marigo (2000, M2K). When the Reimers prescription for mass-loss is adopted, the corresponding mass loss efficiency parameter η is indicated. One can notice that both HG97 and M2K are satisfactorily consistent with the observed data, whereas RV81 is far from reproducing the empirical trend. In particular, the RV81 relation shows a quick divergency of the final mass for increasing initial mass, with the most massive stars being able to build C-O cores up to the Chandrashekhar limit. From P. Marigo (2000) in "The Chemical Evolution of the Milky Way: Stars versus Clusters", ed. F. Matteucci and F. Giovannelli; reproduced here by kind permission of Kluwer Academic Publishers (copy right 2000).

TABLE 2.2. Example of Production Matrix for a star.

	^1H	^2H	^3He	^4He	^7Li	^{12}C	^{13}C	^{14}N	^{16}O	nr	^{20}Ne	^{24}Mg	^{28}Si	^{32}S	^{40}Ca	^{56}Fe	Cu	Zn	Ni
^1H	$1-q_4-LD$																		
^2H		$1-q_3-DL$																	
^3He	w_3	w_2	$1-q_3-w_7$																
^4He	q_4-q_c	q_3-q_c	q_3-q_c	$1-q_c-w_7$	w_4														
^7Li	w_7	w_7	w_7	w_7	$1-w_4$														
^{12}C	$X_{12C}w_c$	$X_{12C}w_c$	$X_{12C}w_c$	$X_{12C}w_c$		A													
^{13}C	$X_{13C}w_c$	$X_{13C}w_c$	$X_{13C}w_c$	$X_{13C}w_c$		C	$1-q_c$												
^{14}N	$X_{14N}w_c$	$X_{14N}w_c$	$X_{14N}w_c$	$X_{14N}w_c$		B		$1-q_c$	B										
^{16}O	$X_{16O}w_c$	$X_{16O}w_c$	$X_{16O}w_c$	$X_{16O}w_c$					D										
nr						w_c	w_c	w_c	w_c	$1-d$									
^{20}Ne	$X_{Ne}w_c$	$X_{Ne}w_c$	$X_{Ne}w_c$	$X_{Ne}w_c$							$1-d$								
^{24}Mg	$X_{Mg}w_c$	$X_{Mg}w_c$	$X_{Mg}w_c$	$X_{Mg}w_c$								$1-d$							
^{28}Si	$X_{Si}w_c$	$X_{Si}w_c$	$X_{Si}w_c$	$X_{Si}w_c$									$1-d$						
^{32}S	$X_{S}w_c$	$X_{S}w_c$	$X_{S}w_c$	$X_{S}w_c$										$1-d$					
^{40}Ca	$X_{Ca}w_c$	$X_{Ca}w_c$	$X_{Ca}w_c$	$X_{Ca}w_c$											$1-d$				
^{56}Fe	$X_{Fe}w_c$	$X_{Fe}w_c$	$X_{Fe}w_c$	$X_{Fe}w_c$												$1-d-E$			
Cu	$X_{Cu}w_c$	$X_{Cu}w_c$	$X_{Cu}w_c$	$X_{Cu}w_c$												w_{Cu_s}	$1-d$		
Zn	$X_{Zn}w_c$	$X_{Zn}w_c$	$X_{Zn}w_c$	$X_{Zn}w_c$												w_{Zn_s}		$1-d$	
Ni	$X_{Ni}w_c$	$X_{Ni}w_c$	$X_{Ni}w_c$	$X_{Ni}w_c$															$1-d$

$LD = w_3 + w_7;$
$DL = w_2 + w_7;$
$A = 1 - q_4 - w_{14N_s} - w_{13C_s};$
$B = q_4 - q_c + w_{14N_s};$
$C = w_{13C_s};$
$D = 1 - q_4 - w_{14N};$
$E = w_{Cu_s} + w_{Zn_s}$

Chapter 3

THE STELLAR BIRTHRATE

3.1   THE PROCESS OF STAR FORMATION

In spite of the fact that the Milky Way has an age of more than 10 Gyr it contains stars of spectral type O which have lifetimes of the order of no more than 3 million years. Therefore, star formation is still an ongoing process in the Galaxy. Such a conclusion is reinforced by the fact that OB stars are almost always surrounded by dense clouds of gas and dust. The first appearance of a hot and brilliant star in a cloud of molecular hydrogen produces immediately an HII region in the area surrounding the star. This HII region has a temperature and density much larger than the ISM and therefore soon or later the larger thermal pressure of the HII region will cause its expansion into the surrounding gas at velocities comparable with the thermal velocity of the HII gas. In this way, the ISM can reach velocities of the order of 10 km sec^{-1}. The expansion of the HII region can induce gravitational instabilities in the nearby regions thus producing the formation of new OB stars. The process of star formation is therefore a self-propagating one. Optical and infrared observations have shown that also the low mass stars appear in giant molecular complexes although they are found preferentially in dark nebulae. A difference between the low and high mass stars is that the former are generally isolated thus indicating that they do not form by means of a hierarchical fragmentation process. On the other hand, massive stars form only in groups and inside the giant molecular clouds. Although most of the physics concerning star formation is still unknown, there is a general consensus in thinking that the formation of *low mass stars* is a primary process in the sense that it implies the fragmentation of clouds into pieces and then the direct collapse of these pieces into stars. On the other hand, the formation of *massive stars* implies a secondary process which occurs as a result of accretion processes in dense environments.

3.1.1   THE FORMATION OF LOW MASS STARS

The youngest low mass stars we know are the so-called T-Tauri stars; they are normally immersed in dense clouds of gas and dust. Their position on the Hertzsprung-Russell (H-R) diagram confirms the fact that these stars are young since they occupy a region above the MS of Pop I stars. This is consistent with the idea that they are still contracting towards the MS.

In a very simple way we review the basic ideas about the formation of a low mass star: let us start with a spherical globule of mass M where the temperature T remains almost constant at 10K during the time when the gas and dust are sufficiently transparent to their own cooling radiation. As this globule starts contracting because of an increase of external pressure P_{ext}, the volume of the cloud begins to decrease. However, as the volume V decreases we expect that self-gravity dominates and that at a certain point it becomes so strong that a further decrease in V does not require any more an increase in P_{ext}. Beyond this point, the globule becomes dynamically unstable to the gravitational collapse. The collapse then proceeds faster in the internal than in the external part of the globule; when the nucleus becomes opaque to radiation, in the sense that it can not radiate away the heating due to compression as fast as its generation by the collapse, the collapse is halted by a strong shock wave. At this point, at the center of the globule there is a hydrostatic object where the Virial theorem holds and it can be called a *protostar*. The initial mass of such a protostar can be as small as $10^{-3} M_{\odot}$ which then slowly accretes material from the surrounding gas and dust. This accretion process takes the protostar to a stellar mass on a timescale of the order of 10^4–10^6 years, depending on the initial conditions which have triggered the collapse. This is an idealized scheme of spherical collapse; in reality one should take into account the rotation and the magnetic fields.

An interesting fact is that the masses of the TTauri stars are very similar to the Jeans mass predicted for the physical conditions of the protostellar clouds. The *Jeans mass* is defined as *the minimum mass that should have a cloud of temperature T and density ρ to be able to collapse*. In particular, the Jeans mass is found by imposing the condition that the gravitational energy of a cloud be larger than its internal energy:

$$|E_{grav}| > E_{int} \tag{3.1}$$

namely:

$$\frac{0.6GM^2}{R} > \frac{1.5R_g TM}{\mu} \tag{3.2}$$

which leads to:

$$M > M_J = (2.5R_g T/\mu G)^{3/2}(4/3\pi\rho)^{-1/2} \tag{3.3}$$

where G is the gravitational constant, R_g is the gas constant and μ is the mean atomic weight per particle. From equation 3.2 we can also define the *Jeans length*, as the maximum radius above which the collapse cannot occur, therefore:

$$R < R_J = 1.5GM/\mu \qquad (3.4)$$

The conditions imposed by the Jeans mass are very restrictive: in fact, a typical HI cloud has a temperature of $\simeq 50$ K, a density of $1.7 \cdot 10^{-23} g/cm^3$ and $\mu = 1$ which leads to a Jeans mass of $M_J \sim 3600 M_\odot$, much larger than the actual masses of the real clouds. Therefore, it is evident that the HI clouds are not gravitationally bound. On the other hand, the typical conditions of a molecular cloud are: $T = 10$ K, $\rho = 1.7 \cdot 10^{-21} g/cm^3$ and $\mu = 2$. This implies a Jeans mass of $M_J \sim 8 M_\odot$, which is of the same order of magnitude as the stellar masses and much smaller than the typical masses of the molecular clouds, namely $10^4 - 10^5 M_\odot$. The fact that the Jeans mass in the molecular clouds is larger than the typical stellar mass ($\simeq 1 M_\odot$) indicates that further fragmentations should take place. In fact, as the collapse proceeds the density increases but the temperature decreases or stays constant (isothermal collapse). This is due to the fact that although the cloud would tend to increase its temperature because of the gravitational compression, the internal energy is partly converted into radiation which can easily escape from the cloud as a result of its low density. Therefore, under these conditions the Jeans mass tends to decrease inducing further fragmentation. The fundamental problem, not yet understood, is how the HI clouds accumulate and compress to form the molecular clouds. The main processes which are thought to be responsible for the formation of the molecular clouds are: cloud-cloud collisions, shock waves associated with the spiral arms and Rayleigh-Taylor instabilities due to the magnetic field. The fragmentation continues until the density is high enough to make the cloud opaque to radiation. This critical point occurs when $\rho \simeq 10^{-13} - 10^{-14} g/cm^3$, which means $M_J(T = 10K) \simeq 0.005 M_\odot$. After this point, a further increase in ρ produces an increase in T and therefore an increase in M_J.

3.1.2 THE FORMATION OF MASSIVE STARS

Given the embryonic nature of our understanding of the star formation process, it is not clear whether there is a different way of forming massive stars relative to low mass stars, but we see indeed some differences between low and high mass stars which we now summarize.

As we have already said, massive stars are seen only in large aggregates of gas and other young stars, and the most massive stars tend to form preferentially in the dense nuclei of large clusters. Unfortunately, the mechanism of formation of massive stars is even less understood than the mechanism of formation of low mass stars. What is clear is that the formation of massive stars should be a more complex process. One reason is that inside the molecular clouds, where massive stars form, there are always turbulent internal motions and it is possible to have even more than one typical Jeans mass. Another reason is that for massive stars the radiation pressure becomes important during the formation process thus inhibiting further accretion of gas onto the forming star, at least in the spherical collapse. Finally, if massive stars form in multiple and closed systems, stellar collisions and mergers can occur and represent a particularly efficient way of creating massive stars without the obstacle of the radiation pressure. Another difference between the formation of low and high mass stars is that the latter take a longer time to form. There is a general consensus in thinking that massive stars form out of the residual gas from the low mass star formation. As a consequence of this, the bulk of the material which condenses into stars goes into low mass stars and a lower and lower quantity of material is available to form the more massive stars. This means that, if the distribution function of stellar masses is a power law (see next section), the exponent should be > 1 for massive stars. Massive stars are concentrated to the spiral arms. This fact may suggest that the spiral structure can be due to the star formation and not vice versa (which is also plausible). Following this line of thought Gerola and Seiden (1978) suggested the *stochastic self-propagating star formation theory* which assumes that the star formation is induced by the massive stars of the previous generation through expansion of the HII regions and supernovae which compress the ISM thus stimulating the formation of new stars. At the same time, star formation is inhibited in regions of recent star formation because of the high temperature of the gas there. The chain starts because of spontaneous star formation resulting from mechanisms such as cloud-cloud collisions. These stars have a finite probability of creating other bright massive stars nearby. Thus a chain-reaction begins; once a star has been created in a given region it becomes harder to create new stars in that same region for a refractory period (a parameter of the model). The other parameters of the model are a spontaneous probability P_{sp} and a stimulated probability P_{st} of star formation. Once these parameters are fixed, the remaining fundamental parameter is the differential rotation $d\omega/dR$ (with ω angular velocity). This feed-back process has some physical justification and observational support, as will be discussed again in Section 3.3. In this theory the spiral structure develops over a period of many galactic rotations as a consequence of the combined effect of the differential rotation and star formation. In particular, the rotation distributes the stars in a spiral structure. Numerical simulations have shown that with this relatively simple theory it is possible to reproduce some spiral patterns although not all of them. For example, it is impossible to reproduce the structure of the *grand design spirals* such as M51.

3.1.3   PRIMORDIAL STAR FORMATION

A crucial uncertainty concerns the characteristic masses of the first generation of stars, usually referred to as Population III stars. The importance of knowing the characteristics of Pop III stars is related to the understanding of the nature of dark matter in galaxy halos and galaxy clusters which represents the dominant constituent of the Universe.

Since the Jeans mass depends on the temperature of the gas, the cooling processes will play a fundamental role in the process of star formation.

When the metals are present, due to the progressive chemical enrichment from dying stellar generations, they represent the major source of cooling and opacity, whereas in the absence of metals, in the primordial gas, the most important coolant is the molecular hydrogen which is present in a small amount since is produced after the decoupling epoch via the reactions:

$$H + e^- \rightarrow H^- + h\nu \tag{3.5}$$

and

$$H^- + H \rightarrow H_2 + e^- \tag{3.6}$$

Although at high densities the H_2 is destroyed via $H_2 + H \rightarrow 3H + e^-$, it is nevertheless an important coolant (Peebles and Dicke, 1968). Besides the H_2, the Lyman-α cooling is an important coolant at temperatures in the range of $10^3 - 10^4 K$. At these temperatures, in the framework of the isothermal primordial fluctuations, the Jeans mass is high but not necessarily very high ($\sim 10 M_\odot$, Silk 1980). Recent more advanced studies taking into account formation and destruction of H_2 as well as detailed hydrodynamics of collapsing clouds, have predicted that the first stars can have masses of the order of $50 M_\odot$ (Abel et al. 1998). When a sufficient quantity of metals is produced these high temperatures, cooling mechanisms are bypassed by molecular and grain cooling which reduce the temperature to the range typical of molecular clouds ($\sim 10 - 100$ K). At this stage one expects star formation to proceed in a way similar to that in the present ISM, allowing for low mass as well as high mass formation.

The limiting metallicity for the formation of low mass stars as well as massive stars is $Z \geq 10^{-5}$ (Silk, 1980). It is interesting to note that stars completely without metals have never been observed, thus suggesting Pop III consisted of massive stars.

Another way of deriving primordial stellar masses is to impose the condition that the free fall time should be equal to the Kelvin-Helmoltz time. The Kelvin-Helmoltz timescale , t_{K-H} is the timescale of quasistatic contraction due to release of gravitational energy (Virial theorem) by stars with no internal energy source, and is the typical timescale for contraction of low mass stars during their pre-main sequence phase decreasing strongly with increasing stellar mass. This regime is identified with TTauri stars which are low mass stars. Given these facts, one sees that imposing the condition that the free fall time is equal to t_{K-H} allows one to find a characteristic stellar mass of the initial mass function.

Since the time of free fall $t_{ff} \propto MT^{-3/2}$ (where T is the temperature) and $t_{K-H} \propto TM^{-2}$, then the characteristic mass will be:

$$M_C = 20(T/10^4 K)^{5/6} M_\odot \tag{3.7}$$

which gives $M_C = 0.4 M_\odot$ for a cold gas, whereas for $T = 10^4 K$ (primordial gas) it gives $M_C = 20 M_\odot$. However, actual protostar models suggest larger values of $M_C = 3 M_\odot$ for material with a solar chemical composition.

3.2 DERIVATION OF THE STELLAR BIRTHRATE

The number of stars formed in the mass interval $m, m + dm$ and in the time interval $t, t + dt$ is the so-called *birthrate function*, $B(m,t)$.

Given the lack of a detailed understanding of the star formation process, the stellar birthrate is usually separated into two independent functions:

$$B(m,t) = \varphi(m)\psi(t) dm dt \tag{3.8}$$

The function $\psi(t)$ is usually known as the star formation rate (SFR), which represents how many solar masses of the ISM are converted into stars per unit area per unit time, and $\varphi(m)$ is the initial mass function (IMF), namely the number of stars ever formed per unit area per unit mass. The sort of indeterminacy problem related to the SFR and the IMF is well known: in order to derive one of these functions one has to know the other. In fact, as we will see in the following sections, what we can do is derive the IMF after having assumed the SFR or vice versa.

3.2.1 DERIVATION OF THE IMF IN THE SOLAR VICINITY

For an extensive discussion concerning the derivation of the IMF in the solar vicinity the reader should consult the reviews of Tinsley (1980) and Scalo (1986); here we recall only the most important steps of this problem.

The current mass distribution of MS stars per unit area, $n(m)$ (the so-called present day mass function, hereafter PDMF), for stars ($0.1 < m/m_\odot \leq 1$) with lifetimes equal to or larger than the age of the Galaxy ($\tau_m \geq t_G$), can be written as:

$$n(m) = \int_0^{t_G} \varphi(m)\psi(t) dt \tag{3.9}$$

where t_G is the galactic lifetime. These stars are therefore still on the MS. If $\varphi(m)$ is constant in time, as it is usually assumed, then:

$$n(m) = \varphi(m) <\psi> t_G \tag{3.10}$$

where $<\psi>$ is the average SFR in the past.

For stars with lifetimes much shorter than the galactic age, namely for $\tau_m \ll t_G$ ($m \geq 2M_\odot$), we see on the MS only those born after the time $t = t_G - \tau_m$. The PDMF will therefore be:

$$n(m) = \int_{t_G - \tau_m}^{t_G} \varphi(m)\psi(t)dt \tag{3.11}$$

Again if $\varphi(m)$ is constant in time it implies:

$$n(m) = \varphi(m)\psi(t_G)\tau_m \tag{3.12}$$

under the assumption that $\psi(t_G) = \psi(t_G - \tau_m)$, where $\psi(t_G)$ is the SFR at the present time, t_G, and τ_m is the lifetime of a star of mass m.

We can apply neither of the previous approximations to the stars in the mass interval $1 - 2\ M_\odot$, therefore $\varphi(m)$ in this mass interval depends on the ratio between the present time SFR and the average total mass of stars ever formed during the galactic lifetime. This quantity can be written as:

$$b(t_G) = \frac{\psi(t_G)}{<\psi_G>} \tag{3.13}$$

It has been shown by Scalo (1986) that a good fit between the two portions of the $\varphi(m)$, namely below $1\ M_\odot$ and above $2\ M_\odot$, requires:

$$0.5 \leq b(t_G) \leq 1.5 \tag{3.14}$$

This condition indicates that the SFR in the solar neighbourhood was almost constant or slightly decreasing in time. This result confirmed what had been suggested already by Twarog (1980) who attempted to determine the SFR history in the solar neighbourhood by measuring directly the age distribution of the local disk stars by isochrone fitting. He studied a large sample of nearby F stars and claimed that the average SFR of these stars over the past 12 Gyr was about 1.5 times the present rate. Therefore, he concluded that the SFR during the lifetime of the disk was a mildly varying function of time and did not exclude the possibility that it was constant. However, Twarog's results depend on the assumed age dependence of the scale height of local field stars, which is not well constrained observationally. He assumed for example that the scale height is constant for stars with ages between 0 and 4 Gyr and then increases linearly with age to a value 6 times as large for the oldest stars. Very probably the scale height has a much stronger age dependence for ages between 0 and 4 Gyr, because the velocity dispersion of local stars has a strong increase with age in this range (Wielen, 1977). Larson (1986) showed that if the SFR of Twarog is recalculated by adopting a scale height increasing proportionally to the velocity dispersion, then the ratio between the past SFR and the present time SFR is about 2.4. Although this value is still quite uncertain, this result does not exclude the possibility of having $b(t_G) = 2.0$.

More recent approaches to reconstructing the star formation history in the Milky Way by means of the age distribution of stars in the solar neighbourhood, which is in principle the most direct way, suggest that the star formation history was not continuous but rather proceeded in bursts with fluctuations of the SFR larger than a factor 2-3 on timescales less than 0.2-1 Gyr (Rocha-Pinto et al. 1999). However, in the study of the chemical evolution of the Galaxy and galaxies in general, what matters is the integral over time of the SFR rather than the SFR itself, since the gas chemical abundances derive from the integration of the star formation history over the galactic lifetime. Therefore, if the average SFR in the past has not varied much relative to the present time SFR, and the fluctuations in the star formation are not too large, one would not expect large differences in the predicted chemical evolution. This suggestion still awaits to be tested by chemical evolution models.

For $m < 0.1 M_\odot$ we do not know $n(m)$ and therefore cannot derive $\varphi(m)$. The determination of $n(m)$ at the bottom of the MS is, in fact, subject to major uncertainties due to either the very low intrinsic luminosity of these stars or to their rapidly varying Mass-Luminosity (ML) relation (see D'Antona, 1995).

It is clear from the previous equations that, once the PDMF is derived observationally (see 3.2.2), either the IMF can be determined if the SFR is known or the SFR can be derived if the IMF is assumed. Unfortunately, a choice of an IMF combined with an arbitrary SFR will not, in general, satisfy the fundamental constraint represented by the observed number of MS stars. Therefore, it is very important to reproduce the PDMF in models of galactic chemical evolution.

3.2.2 DERIVATION OF THE PDMF IN THE SOLAR VICINITY

Information on the PDMF is usually obtained by counting stars as a function of their absolute magnitude and obtaining the luminosity function $\frac{dN}{dM_v}$. This function is then corrected to account for the presence of post-MS stars in each interval of absolute magnitude. This implies the knowledge of colour and spectral type of the stars. Then, the star counts are transformed into the average number of stars per pc^2 at a given galactocentric distance, since stars of different mass are differently distributed perpendicularly to the galactic plane (the scale height decreases with increasing mass). At this point the ML relation for stars in the MS phase allows one to pass from absolute magnitude to mass.

Therefore, we define the PDMF as:

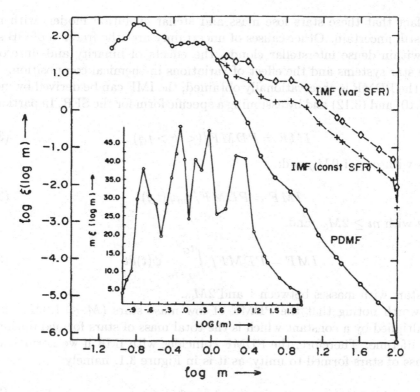

Figure 3.1. PDMF and IMF under different assumptions about the SFR (constant and variable). Both the PDMF and the IMF here are defined in units of number per *logm* per pc^2. Note that the slope of the IMF in the high mass range is not sensitive to the degree of variability of the SFR, although the overall slope changes. In the insert is shown the IMF derived by adopting a particular law for the SFR including a dependence on the metallicity of the ISM. From Rana 1991, A.R.A.A Vol. 29, 129; reprinted, with permission, from the Annual Review of Astronomy and Astrophysics (copy right 1991).

$$\phi(m) = \frac{dM_v}{dm} 2H(M_v) f_{ms}(M_v) \frac{dN}{dM_v} \tag{3.15}$$

where $\frac{dM_v}{dm}$ is the derivative of the absolute magnitude-mass relation, $H(M_v)$ is the correction for the increase of the scale heights of the stars in the faint part of the luminosity function with a given absolute magnitude and $f_{ms}(M_v)$ represents the corrections for the percentage of evolved stars at a given absolute magnitude. These last two factors depend on the history of the galactic disk. For example, the ratio between the number of the evolved and MS stars is higher if the SFR was higher in the past. The largest uncertainty in deriving the PDMF of low mass stars resides in the scale heights, while for high mass stars the main uncertainty is due to the mass-absolute magnitude relation and luminosity function. For example, concerning the luminosity function of massive stars, one should account for the fact that these stars lose mass and stellar evolution models with mass loss are still uncertain. Other causes of uncertainty are: the fraction of OB stars hidden within dense interstellar clouds, the effects of binarity and unresolved multiple star systems and the effect of variations in chemical composition.

Once the PDMF is observationally obtained, the IMF can be derived by means of eqs (3.10) and (3.12) and by assuming a specific form for the SFR. In particular:

$$IMF = PDMF/(<\psi> t_G) \qquad (3.16)$$

for stars with $m \leq 1.0 M_\odot$ and:

$$IMF = PDMF/\tau_m \psi(t_G) \qquad (3.17)$$

for stars with $m \geq 2 M_\odot$, and:

$$IMF = PDMF/ \int_{t_G - \tau_m}^{t_G} \psi(t)dt \qquad (3.18)$$

for the stars with masses between 1 and $2M_\odot$.

It is worth noting that the PDMF of low mass stars ($M \leq 1.0 M_\odot$) is the IMF multiplied by a constant which is the total mass of stars formed during the galactic lifetime. Therefore, the PDMF coincides with IMF if we normalize the total mass of stars formed to unity, as it is in Figure 3.1, namely:

$$<\psi> t_G = 1 \qquad (3.19)$$

On the other hand, the PDMF in the other mass ranges never coincides with the IMF because the stars in these ranges had time to be born and die several times during the galactic history and either the knowledge of the stellar lifetimes or the star formation history is required.

3.2.3   THE INITIAL MASS FUNCTION

It has been shown by many authors that the IMF in the solar neighbourhood, obtained in the way described in the previous sections, is best approximated by a power law and is assumed to be constant in space and time.

The IMF is generally approximated as:

$$\varphi(m) = Am^{-(1+x)} \qquad (3.20)$$

where A is a normalization constant. For the purpose of calculating the chemical evolution of the Galaxy we generally obtain A by normalizing $\varphi(m)$ to unity as is normally done with probability functions:

$$\int_0^\infty m\varphi(m)dm = 1 \qquad (3.21)$$

The simplest and still most popular formulation for the IMF is that by Salpeter (1955), with the exponent of the power law $x = 1.35$ for masses below $10M_\odot$. In models of chemical evolution of galaxies the Salpeter index is often used over the whole range of stellar masses which is usually defined to be 0.1-$100M_\odot$. The lower mass limit of this range is imposed observationally by the intrinsic faintness of stars with $M_v > 16^m$, whereas the upper limit is quite uncertain. Spectroscopic analyses of hot luminous stars seem to indicate an upper mass limit of $\simeq 100M_\odot$ (M92).

The Salpeter IMF, when defined in the above mass range, can be written as:

$$m\varphi(m) = 0.17m^{-1.35} \tag{3.22}$$

Miller and Scalo (1979) re-derived the IMF for the solar vicinity, for a range of SFR histories consistent with the continuity constraint (namely the continuity in the mass range 1-$2M_\odot$), and showed that a good approximation for the IMF is given by a log-normal mass distribution. They suggested that the slope of the IMF x at a given mass is:

$$x = \frac{d\varphi(logm)}{dlogm} = -(1 + logm) \tag{3.23}$$

This IMF is continuously increasing down to $m \sim 0.1M_\odot$ at variance with a later suggestion by Scalo (1986) about the IMF having a peak at around $0.2M_\odot$. This suggestion was based on the flattening or dip observed in the luminosity function of MS solar neighbourhood stars at $M_v = 6 - 9$, interpreted by Scalo to be due to a corresponding feature in the IMF. In the following, this interpretation has been questioned by D'Antona and Mazzitelli (1986) who correctly attributed this feature in the luminosity function of stars to the non-linear behaviour of the mass-M_v relation. However, the main point of the work of Scalo was to demonstrate that the field star IMF does not exhibit the log-normal form found by Miller and Scalo and that a power law should be preferred. But he also showed that a single power law is not a good representation of the IMF (see Figure 3.2) as already suggested by Tinsley (1980).

Therefore, since the work of Scalo power laws divided into various segments have been preferred to the Salpeter law. Most of the disagreement between different authors still seems to reside in the low mass end of the IMF, namely on whether the IMF flattens, decreases or increases at the bottom of the MS. The reasons for this are the major uncertainties due to the very low intrinsic luminosity of these stars and their rapidly varying ML relation.

In particular, Scalo (1998) presented a survey of results concerning the IMF based on star counts and concluded that the data suggest a three-segment power law, where the slope for intermediate mass stars is steeper than for massive stars, with the following power indices:

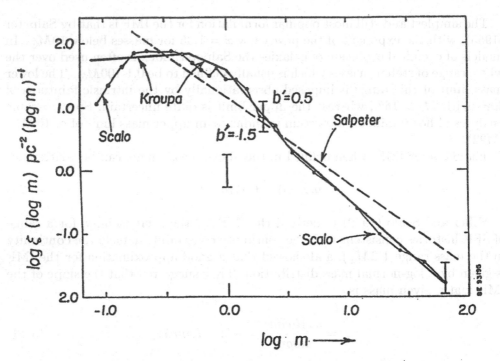

Figure 3.2. Local IMF (here indicated as $\xi(logm)$) after Scalo (1986) with $b' = 12b(t_G)/t_G = 1$ (points joined by thin lines), Kroupa et al. (1990) (thick lines below m=1) and two additional power-law segments approximating Scalo's IMF. Salpeter's law is shown by a broken line. All the IMFs are normalized to a total number of stars ever born of $37M_\odot pc^{-2}$ between the mass limits m=0.1 and m=100$M_\odot$. From Pagel 1997, "Nucleosynthesis and Chemical Evolution of Galaxies"; reproduced by kind permission of B.E.J. Pagel and the Cambridge University Press (copy right 1997).

$$x_1 = 0.2 \pm 0.3 \qquad (3.24)$$

for $0.1 - 1M_\odot$,

$$x_2 = 1.7 \pm 0.5 \qquad (3.25)$$

for $1 - 10M_\odot$,

$$x_3 = 1.3 \pm 0.5 \qquad (3.26)$$

for $10 - 100M_\odot$.

A plausible physical explanation for this kind of IMF is not yet available thus representing a challenge to theorists.

It is worth noting that all the proposed IMFs were derived for the solar neighbourhood domain and we do not know if the IMF is universal or if it varies from galaxy to galaxy or even inside the Galaxy. In past years several authors have proposed variations of the IMF both in space and time but there are still some unsolved problems related to the possible variations of the IMF. The main problem concerning the variation of the IMF in space is the fact that most of the proposed variations tend to destroy the abundance gradient observed along the galactic disk (see chapter 1 and chapter 6). In fact, the most common assumption is based on the Jeans mass being larger in an ambience with a lower metal content, due to the higher temperature which is a consequence of the less effective cooling. This means that the most external regions of the disk would form more massive stars than the innermost regions, with the consequence of preventing the formation of a negative abundance gradient, since massive stars are the responsible for the production of the bulk of metals.

Only the opposite case would work, namely to assume that the formation of low mass stars is favored in the most external regions of the disk. However, this solution does not seem to have many physical justifications. Concerning the variation of the IMF in time, the most common suggestion has been that the IMF was top-heavy at early times on the basis of the following arguments:

- The first reason for proposing that the IMF must have been dominated by massive stars at early times, besides the Jeans mass argument discussed before, is the fact that the standard Big Bang cosmology predicts that the first stars contained no heavy elements and yet no metal-free stars have ever been found.

- Another reason is the necessity of solving the "G-dwarf problem", namely the paucity of metal poor stars in the solar neighbourhood when compared with the predictions from the simple model of chemical evolution (see chapter 5).

- Chemo-photometric evolution models for elliptical galaxies, aimed at reproducing the colors and abundances in the stellar populations and in the intra-cluster medium (ICM), indicate that the IMF in these systems should contain more massive stars than the local one, although chemo-photometric models still contain many uncertainties which do not allow one to draw firm conclusions on this point.

- There are also suggestions that starburst galaxies should have an IMF either with an excess of massive stars or with a deficiency of low mass stars. Again, these suggestions mostly arise from models of population synthesis which suffer from many uncertainties and do not give a unique solution. On the other hand, from the point of view of the chemical evolution of starburst galaxies there is no compelling evidence that the IMF should be different from a normal Salpeter-like IMF.

- The fact that the standard Big Bang cosmology suggests that a large fraction of the dark matter in the Universe should be in the form of baryonic matter indicates that perhaps most of the stellar remnants were formed from the early stellar generations which should therefore contain a high percentage of massive stars.

The last suggestion concerning a top-heavy IMF at early times is from Larson (1998) who claimed that the IMF has a universal Salpeter-like form at the upper end, but flattens below a characteristic stellar mass which may vary with time. In particular, he suggested:

$$\frac{dN}{dlogm} \propto (1 + m/m_1)^{-1.35} \qquad (3.27)$$

This IMF has a logarithmic slope $x = 1.35(1 + m_1/m)^{-1}$; it approaches the Salpeter form at large masses but becomes asymptotically flat with $x = 0$ at the low mass end. The mass scale m_1 may be variable and perhaps be associated to the Jeans mass being larger at early times.

However, before drawing any firm conclusion on which IMF is the best for the solar neighbourhood and the whole Galaxy, these IMFs should be tested in chemical evolution models to see if they can reproduce the majority of the observational constraints. As we will see in chapter 6, one of the most crucial tests for a variable IMF are the abundance gradients along the galactic disk.

3.2.4 DERIVATION OF THE SFR

In order to derive $\psi(t_G)$ the procedure to follow is to assume a priori an IMF and then integrate the equations which give the PDMF with respect to the mass. In this way one can obtain the present time SFR for the solar neighbourhood region, which is the only region where we know the PDMF.

Miller and Scalo (1979) followed this procedure and derived the present time SFR, once a reasonable IMF was assumed. Timmes et al. (1995) have summarized all the available estimates of the present time local SFR including that of Miller and Scalo and suggested:

$$\psi(t_G) \sim 2 - 10 M_\odot pc^{-2} Gyr^{-1} \qquad (3.28)$$

As we have already discussed, it is very difficult to derive the temporal history of the SFR in the solar vicinity; constraints on the time dependence of the SFR are provided by the fit of the continuity constraint on the IMF and by the observed properties of stellar populations.

In order to derive the present time SFR at different galactocentric distances and in external galaxies one should use different tracers. Most direct tracers of SFR in galaxies are sensitive only to the luminosities of massive stars, above 5-10 $M_\odot$. Unfortunately, for typical IMFs these massive stars represent only 5-20% of the total mass of the stellar population, so deriving a total SFR involves a large (factor 5-20) extrapolation over the stellar component which is actually observed.

The most common tracers of star formation are:

- counts of the supergiants which are luminous enough to be seen also in nearby galaxies under the assumption that their number is proportional to the SFR.

- Otherwise one can measure the H_α and H_β flux from HII regions, which are ionized by young and hot stars, and assume that such flux is proportional to the SFR. In Kennicutt (1998a), where the reader can find an exhaustive review on the derivation of the SFR in galaxies, the following relation is suggested:

$$SFR(M_\odot yr^{-1}) = 7.9 \cdot 10^{-42} L_{H_\alpha}(erg sec^{-1}) \qquad (3.29)$$

for the total SFR, measured in a sample of field spiral and irregular galaxies, after adopting the Salpeter IMF over the mass range 0.1-$100 M_\odot$

- From the integrated UBV colours and spectra of galaxies one can estimate the relative proportions of young and old stars to have an idea of the ratio between the present time SFR and the average SFR in the past. However, this procedure requires good population synthesis models and these models still contain many uncertainties, mainly because of the unsettled problem of the age-metallicity degeneracy, namely the fact that variations in age and metallicity produce the same effects on the integrated colors and spectra of galaxies, thus preventing us from disentangling the two effects correctly.

- The frequency of type II SNe as well as the distribution of SN remnants and pulsars can be used as tracers of the SFR.

- The radio emission from HII regions can also act as a tracer of the SFR.

- The UV ultraviolet continuum and the infrared continuum (star forming regions are surrounded by dust) are also connected to the SFR. For example, Donas et al. (1987) used UV data ($\sim 2000 \text{Å}$) to estimate the SFR.

With the following choice of the IMF:

$$\varphi(m) \propto m^{-1.6} \qquad (3.30)$$

for $m \leq 1.8 M_\odot$ and

$$\varphi(m) \propto m^{-3} \qquad (3.31)$$

for $m > 1.8 M_\odot$, they derived:

$$SFR(M_\odot yr^{-1}) = 0.9 \cdot 10^{-6} \frac{L(UV)}{L_{bol\odot}} \qquad (3.32)$$

- Finally, the SFR can be derived from the distribution of molecular clouds (see Rana 1991).

From these studies it has been inferred that most of the star formation activity in the Galaxy occurs in the 4 kpc ring, roughly half way between the Sun and the galactic center. There is also a strong star formation activity around the center itself but little in the region in between the center and the 4 kpc ring, the main reason being probably a shortage of gas which could be due to dynamical effects produced by the bar which seems to be present in the galactic bulge. The distribution of the SFR as a function of the galactocentric distance is shown in Figure 3.3 where data from the distribution of SN remnants and pulsars, from Lyman-continuum photons and from molecular gas, are collected. The $SFR(R)$ is normalized to the $SFR(R_\odot)$ at the solar radius.

It is also possible to estimate the SFR in the past; in particular, as shown by Gallagher et al. (1984), one can estimate the SFR over the lifetime of a galaxy from the dynamical mass contained within the optical radius corrected for diffuse gas. This mass, in fact, is composed of mostly low mass stars. The SFR over the past 3-4 Gyr can be derived from the blue luminosity produced by MS stars of masses between 1 and 2.5 $M_\odot$.

3.3   THE PARAMETRIZATIONS OF THE SFR

Considering the fact that we know very little about the physical process of star formation, it is convenient to describe the SFR by means of a simple law, involving parameters such as the surface gas density, the total surface mass density or galactic rotation constants, all of which probably play a role in the star formation process. Unfortunately, there is no clear basis for such simple laws and other parameters like spiral shocks, magnetic fields and cloud-cloud collisions may play a fundamental role.

Lynden-Bell (1977) pointed out that a proper expression for the SFR would contain so many unknown parameters as to be useless. In particular, he wrote:

$$\psi = \psi(\rho_{gas}, c_s, \omega_s, V_T/c_s, \Omega, A, |B|, Z, \rho_*) \qquad (3.33)$$

where ρ_{gas} is the volume gas density, c_s is the sound speed, ω_s is the shock frequency, V_T/c_s is the shock strength, Ω is the gas rotation, A is the Oort constant measuring the shearing rate, $|B|$ is the magnetic field strength, Z is the gas metallicity and ρ_* is the background stellar density. Therefore, we are forced to

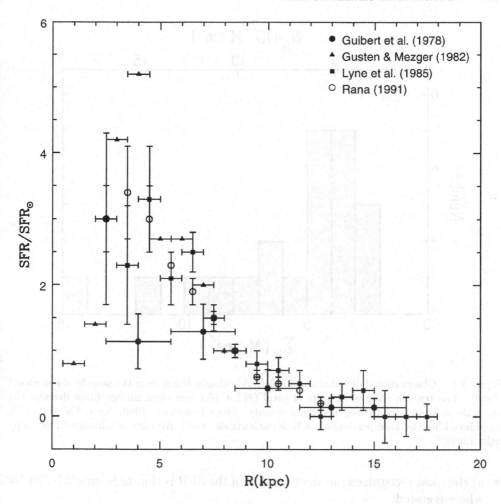

Figure 3.3. The distribution of the ratio $SFR(R)/SFR(R_\odot)$ as a function of the galactocentric distance R in our Galaxy. The SFR has been derived by the distribution of SN remnants (Guibert et al. 1978), from the distribution of pulsars (Lyne et al. 1985), from the Lyman-continuum photons (Güsten and Mezger 1982) and from the distribution of molecular clouds (Rana 1991). From Portinari 1998, PhD Thesis University of Padua; reproduced by kind permission of L. Portinari.

parametrize the SFR guided by the above ideas on the relevant factors influencing the star formation process, with the aim of testing how galactic evolution is affected by these various factors.

Various parametrizations of the SFR have been derived up to now by looking at a possible correlation between the SFR and the gas density. These include :

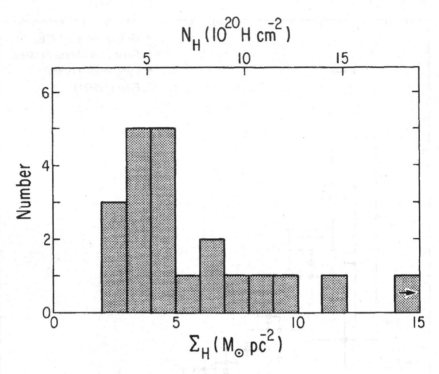

Figure 3.4. Observational distribution of threshold column densities in the sample of Kennicutt (1989). The bottom scale is in units of total $(HI + H_2)$ hydrogen surface mass density, the top scale is in units of hydrogen column density. From Kennicutt 1989, Ap.J. Vol. 344, 685; reproduced here by kind permission of R. Kennicutt and the University of Chicago Press (copyright 1989).

- a) the most recognized parametrization of the SFR is that of Schmidt (1959;1963) who suggested:

$$\psi(t) \propto \rho_{gas}^{k} \qquad (3.34)$$

with $k = 1 - 2$ and ρ_{gas} being the volume gas density. He measured the space density of stars in different regions of the Galaxy in relation to the number density of neutral hydrogen (HI) measured by means of the 21 cm emission line. He did not consider molecular hydrogen since no measures of H_2 were available at that time.

- b) a SFR proportional to some power, k (between 1 and 2), of the total surface gas density $[1.4(HI + H_2)]$, where the factor 1.4 takes into account the contribution of He):

$$\psi(t) \propto \sigma_{gas}^k \qquad (3.35)$$

This formulation is similar to Schmidt's original suggestion a). In principle, Schmidt's formulation of the SFR and b) are equivalent when $k = 1$, whatever the scale height of gas, but they are also equivalent for any value of the exponent as long as the scale height of gas is constant with galactocentric radii. The advantage of the formulation in terms of the surface gas density lies in the fact that it does not require the computation of the scale height of the gas layer, which is poorly known as is the evolution of this scale height with time.

In a series of papers Kennicutt (1983;1989;1998a,b) tried to assess the dependence of massive star formation on the surface gas density in disk galaxies, by comparing the H_α emission with the data on the distribution of HI and CO, and found that in dense regions the SFR can be well represented by a law, like eq. (3.35), with $k = 1.4 \pm 0.15$. He also showed that this law breaks down at densities below a critical threshold value and it becomes much steeper than the above Schmidt law at densities near the threshold. The existence of such a threshold in the star formation appears to be associated with the onset of large-scale gravitational instabilities occurring in the gas of disks. In fact, below a certain critical density the gas disk is stable against the growth of large-scale density perturbations, and consequently one would expect clouds to grow and star formation to be strongly suppressed. The observed threshold densities agree with the threshold densities predicted by simple single fluid disk stability models (Toomre 1964). The value of this critical density is not unique since it is found to vary from galaxy to galaxy as shown in Figure 3.4, where the distribution of threshold column densities in the galaxy sample of Kennicutt is reported.

The adoption of a threshold in the SFR in chemical evolution models can produce very different results from those without such a threshold (see next chapters). Another very interesting aspect of the study of Kennicutt is his finding that there is a better correlation between SFR and total gas $(HI+H_2)$ density rather than between SFR and molecular gas density (H_2), as one would think, given the fact that star formation takes place in molecular clouds! The correlation between the SFR and the total surface gas density for spiral and starburst galaxies is shown in Figure 3.5. The best fit to these data implies:

$$SFR = (2.5 \pm 0.7) \cdot 10^{-4} \left(\frac{\sigma_{gas}}{1 M_\odot pc^{-2}}\right)^{1.4 \pm 0.15} M_\odot kpc^{-2} yr^{-1}. \qquad (3.36)$$

Perhaps the coupling between molecular gas and the SFR is a strictly local process, which is obliterated when we study the SFR on very large scales, or

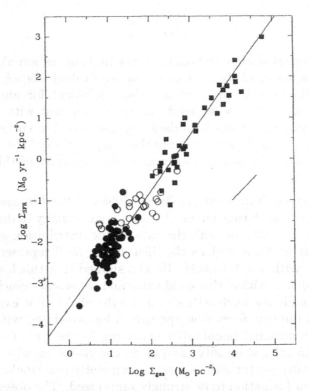

Figure 3.5. Observational correlation between the SFR and the surface gas density of the total gas for normal disks (circles) and starburst galaxies (squares). The line represents a least square fit with $k = 1.4$. The short, diagonal line shows the effect of changing the scaling radius by a factor of 2. From Kennicutt 1998, Ap.J. Vol. 498, 541; reproduced here by kind permission of R. Kennicutt and the University of Chicago Press (copy right 1998).

perhaps variations in the CO/H_2 conversion ratio are introducing a spurious dispersion in the correlation between SFR and molecular hydrogen.

- c) A SFR for the Galaxy proportional to the surface mass density of molecular hydrogen but with a dependence on the metallicity was proposed by Rana and Wilkinson (1986):

$$\psi(t) \propto \sigma_{H_2}/\sigma_{(HI+H_2)} \propto Z^{1.3} \qquad (3.37)$$

where Z represents the metallicity, in particular the oxygen abundance derived from the HII regions. Tosi and Diaz (1990) studied a similar SFR by means of models of chemical evolution and concluded that this particular formulation of the SFR produces results in poorer agreement with the properties of disk galaxies than models assuming an exponentially decreasing SFR.

- d) A more complicated formula for the SFR can be obtained by introducing a dependence on the total surface mass density as well as the surface gas density:

$$\psi(t) \propto \sigma_{gas}^m \sigma^n \qquad (3.38)$$

where σ is the total surface mass density. This kind of formulation was first suggested by Talbot and Arnett (1975) and more recently by Dopita and Ryder (1994) who claimed that a satisfactory agreement with the observed properties of disk galaxies, shown in Figure 3.6, can be obtained with a law of this kind with $1.5 < m + n < 2.5$. In particular, the proposed star formation law is:

$$\psi(t) \propto \sigma_{gas}^{5/3} \sigma^{1/3} \qquad (3.39)$$

The physical basis for such an assumption is the existence of a feed-back mechanism regulating star formation: on one side there is the input of energy into the ISM provided by supernovae, massive stars and HII regions, on the other side there is the timescale for the gas to cool and reach the conditions for fragmentation and therefore star formation to take place. In other words, the self-regulating mechanism of star formation implies that when stars form in a given region and inject energy into the surrounding ISM, subsequent star formation in that region is inhibited until the gas cools again. As a consequence of this, the large scale star formation process is regulated by the self-gravity of the gas (namely the total surface mass density) which produces stars at a rate proportional to the average gas density.

In a scenario where the disk forms by infall of gas which cools and settles into a thin disk, the rate of gas accumulation and star formation is controlled by the rate at which the turbulent energy is dissipated eventually reaching an equilibrium condition. If one assumes that the timescale of energy dissipation is of the order of the period of vertical oscillations through the galactic plane, then the SFR will depend, besides the surface gas density, on the total surface mass density in the disk. A consequence of the self-regulating mechanism of star formation is again the existence of a threshold in the gas density. Such a threshold can be identified with the virial gas density above which star formation begins on a scale equal to the Jeans length.

The self-regulating mechanism of star formation seems to be indicated for spiral and irregular galaxies but it probably fails in spheroidal systems such as the galactic bulge or elliptical galaxies. In fact, as pointed out by Elmegreen (1999), an important difference is that the bulge potential well is too deep to have allowed self-regulation or blowout of the gas by the pressures from young stars and supernovae. As a result, the bulge formation should have been at a maximum rate with most of the gas converted into stars in few dynamical timescales, in other words the bulge should have suffered a starburst and this agrees with other considerations on the bulge stars that we will discuss in chapter 6.

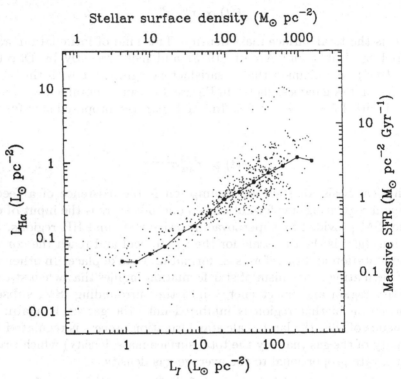

Figure 3.6. Observational correlation between the I-band and the H_α surface brightness of galactic disks. From Dopita and Ryder 1994, Ap.J. Vol. 430, 163; reproduced here by kind permission of M. Dopita and the University of Chicago Press (copy right 1994).

- e) A similar approach has been based on the analysis of induced and spontaneous star formation (Franco and Shore 1984; Shore and Ferrini 1995), which leads again to a self-regulating mechanism for star formation similar to the one proposed by Talbot and Arnett. This SFR also depends on both surface gas density and total surface mass density, where the exponents of the gas density and mass density are fixed by the evolutionary stage of the interacting superbubbles created by the OB star associations. In particular, starting from a Schmidt law, the exponent in this case is $k = 59/40$ (see Matteucci et al. 1989). The SFR expressed in surface densities is:

$$\psi(t) = 0.02\sigma^{19/40}\sigma_{gas}^{59/40} M_\odot pc^{-2} Gyr^{-1}. \tag{3.40}$$

In general, star formation rates depending on both surface gas density and total surface mass density reproduce the main features of the galactic disk better than the simple Schmidt law, as discussed in chapter 6.

- f) A bimodal star formation rate, with a "high" mode of star formation exponentially decreasing with time and a "low" mode of star formation independent of time (Larson, 1986; Wyse and Silk 1987; Vangioni-Flam and Audouze 1988) have been proposed, but it has been shown that none of them can satisfactorily reproduce all of the observational constraints of the Galaxy (Tosi, 1988; François et al. 1990) and introduces additional free parameters. Moreover, these bimodal SFRs tend to overestimate the predicted stellar remnants compared to what is observed (Larson, 1991).

- g) Another approach is the Schmidt law of star formation which depends only on the cold gas, as proposed by Burkert and Hensler (1988) in their chemo-dynamical models. The assumed SFR is:

$$\psi(t) = \alpha_{SF}\eta_{SF}(T)\rho_{CM}^2 \tag{3.41}$$

where the coefficient $\eta_{SF}(T)$ is a temperature dependent factor and ρ_{CM} is the density of the cloud medium. The properties of such star formation rate have been further studied in Köppen et al. (1995), who suggested that feedback between gas and stars naturally leads to a quadratic dependence of the star formation rate on the gas density, independent of the other assumptions on the stellar birthrate function.

However, this formulation also contains several free parameters, and this is unavoidable given the lack of knowledge of the physical processes which influence the formation of stars.

- h) In galactic chemical evolution models an exponentially decreasing SFR is often used:

$$\psi(t) \propto e^{-t/\tau} \tag{3.42}$$

where τ is a free parameter which should be between 5 and 15 Gyr to allow for model predictions to be in agreement with observational constraints (Tosi, 1988).

- i) Another important SFR proposed in the literature assumes a dependence on the angular frequency of the gas in the disk as well as on the frequency of the spiral pattern, in addition to the dependence on the surface gas density. This suggestion is based on the idea that stars are formed in the galactic disk when the ISM with angular frequency $\Omega_g(R_G)$ (where R_G is the galactocentric distance) is periodically compressed by the passage of the spiral pattern, having a frequency $\Omega_P = const. << \Omega_g(R_G)$. This leads to $SFR \propto (\Omega_g(R_G) - \Omega_P)$ and for a flat rotation curve to $SFR \propto R_G^{-1}$ (Wyse and Silk, 1989).

Recently, Boissier and Prantzos (1999) proposed the following SFR, based on the above considerations:

$$\psi(t, R_G) = 0.1(R_G/R_\odot)^{-1}\sigma_{gas}^{1.5}M_\odot pc^{-2}Gyr^{-1} \qquad (3.43)$$

where $R_\odot = 8.0$ kpc is the assumed distance of the Sun from the galactic center. From the point of view of models of galactic evolution, this particular formulation seems to account for the abundance gradients as well as the gas and star formation rate distributions (see chapter 6). From the observational point of view, Kennicutt (1998b) found a correlation between the observed SFR and the ratio of the surface gas density to the local dynamical timescale, $\frac{\sigma_{gas}}{\tau_{dyn}}$. In particular, from this correlation one can derive the following parametrization, known as *Kennicutt's law*:

$$\psi(t) \propto \frac{\sigma_{gas}}{\tau_{dyn}} \propto \sigma_{gas}\Omega_g = 0.017\sigma_{gas}\Omega_g \qquad (3.44)$$

where Ω_g is the angular rotation speed of the gas. This parametrization derives from the same data of Figure 3.5 and, in principle, offers an equally good parametrization of the global SFR in galaxies.

In most of the SFR parametrizations discussed above the main physical assumptions rest on the functional form of the assumed law and on the exponent attributed to the gas density and mass density. We did not mention the proportionality constant which should be present in all the star formation laws mentioned above. Such a proportionality constant has the physical meaning of the *efficiency of star formation* and is expressed in units of $time^{-1}$. We will refer to it as to ν_{SF}. The inverse of ν_{SF} is the timescale for the complete gas consumption in the considered galaxy or galactic region and is equivalent to the timescale of star formation as defined in chapter 5. Generally, in galactic chemical evolution models this efficiency is a free parameter which is assumed to be constant and is chosen in order to reproduce the present time observed SFR.

In conclusion, all of the various parametrizations of the SFR need to be tested in chemical evolution models by attempting to reproduce the maximum number of observational constraints. In this way one can decide which SFR best reproduces the reality, a matter to be discussed in chapter 6.

Chapter 4

GAS FLOWS

Gas flows in and out of galaxies are fundamental ingredients for studying their chemical evolution, since they are required to explain several important features in galaxies and galaxy clusters. These features include the metallicity distribution of G-dwarfs in the solar neighbourhood, the abundance gradients along the galactic disk as well as the heavy element abundances measured in the ICM. We will discuss here the infall of gas onto disks, the radial flows along disks and galactic winds.

4.1　INFALL OF GAS ONTO THE GALAXY

Oort in 1970 first discussed the possibility of matter infalling onto the disks of spiral galaxies. He envisioned that the penetration into the Galaxy of extragalactic neutral gas clouds with very high velocities (VHVC; $|V| > 140$ km sec^{-1}) can trigger the formation of high velocity clouds (HVC; $80 \leq |V|$ km sec$^{-1} \leq 140$) when they interact with galactic matter. He suggested that the present time infall rate onto the Galaxy should be of the order of $1 M_\odot yr^{-1}$.

Mirabel and Morras (1984; 1990) presented observations at 21 cm of HI in the direction of the galactic anticenter showing that a stream of VHVC has reached the outer Galaxy and is interacting with galactic matter. Their HI survey provided evidence for the accretion of gas onto the Galaxy at very high velocities: more than 99% of the VHVC in the direction of the galactic anticenter and 84% of the VHVC in the inner Galaxy have negative (approaching) velocities.

It should be noted that the computation of the infall rate onto the Galaxy depends on several unknown factors such as the distance to the clouds, the motion of the objects in the plane of the sky and the actual distribution of infalling gas over the whole sky and therefore the observational estimates of the infall rate should be regarded as still uncertain. Mirabel and Morras derived, from a survey of VHVC, a total infall rate onto the galactic disk of 0.2 - $0.5 M_\odot yr^{-1}$.

The origin of VHVC is not known but very probably they are made of extragalactic gas. On the other hand, HVC could have a galactic origin, as originally

135

proposed by Oort, and could have been set in motion by the interaction with the VHVC. The HVC could subsequently cause large-scale disturbances such as HI supershells, as suggested by Tenorio-Tagle (1980).

Analyses of intermediate, high and very high velocity clouds (IVC, HVC, VHVC) by means of UV, optical and radio measurements along the line of sight of globular clusters (e.g. De Boer and Savage, 1984) and halo stars (Songaila et al. 1988; Danly 1989) have shown that no cloud has a height z above the galactic plane lower than 300 pc. Therefore, this seems to rule out a local origin for all of these clouds, otherwise they should also be found at lower galactic latitudes and show $\sim 50\%$ of positive velocities. Danly (1989) also found that whilst the northern galactic hemisphere shows many approaching clouds, the southern hemisphere is almost empty. This could be an indication of the intergalactic origin of the gas, captured only by the leading face of the Galaxy in its motion towards Virgo. The rate of gas infall extrapolated from these observations is around $1 - 2 M_\odot$ yr^{-1}, obviously larger than that estimated from solely the VHVC.

In order to ascertain the origin of all these clouds it would be very important to measure their chemical composition (i.e. a roughly solar metallicity would prove a local origin), but the available data are not good enough to draw firm conclusions.

Braun and Burton (1999) have identified a particular class of HVC, which might represent a homogeneous subsample of these objects, in a single physical state. These clouds are compact (CHVC) and apparently isolated; they possess an infalling velocity of the order of 100 km sec^{-1} in the Local Group reference frame. The interesting aspect of this study is that it suggests that the CHVC are probably not the consequence of a galactic fountain and that they rather have an extragalactic origin. They could represent examples of collapsed pristine gas with very little internal star formation and enrichment, the building blocks of galaxies in a hierarchical structure formation scenario (see also Blitz et al. 1999).

In the past, the existence of infall onto the Galaxy has been challenged by Ostriker who pointed out that if the infalling gas stopped abruptly when hitting the disk, this energy should be radiated in the X-rays, at variance with the observed X-ray background. The temperature of the collisionally heated gas would be in the range of $10^6 - 10^7$ K for velocities in the range of 100-300 km sec^{-1}, which would correspond to a radiation in the range $0.25 - 2.5$ keV. On the other hand, Mirabel and Morras argued that these soft X-ray photons would be absorbed by the interstellar HI after traveling a few parsecs through the galactic disk, and that there is no chance of detecting the collisionally ionized gas.

From a theoretical point of view, the existence of infall of gas onto the galactic disk was claimed as a natural consequence of a realistic galaxy formation process from extended halos, and to solve the G-dwarf problem in the solar vicinity (Larson 1972; Lynden-Bell 1975). Infall is also desirable to prevent gas consumption in spirals in times shorter than their ages (Tinsley, 1980 and next chapter). Larson (1991) has shown that the infall rate of Mirabel and Morras, if transformed into a rate per unit area, by assuming (although it is very uncertain) that the infall rate is uniform over a disk of radius 15 kpc, gives $0.3\text{-}0.7 M_\odot pc^{-2} Gyr^{-1}$, which is only a fraction between 0.16 and 0.4 times the local gas depletion rate of $1.8 M_\odot pc^{-2} Gyr^{-1}$. The local gas depletion rate can be estimated by assuming that the average SFR in the last 12 Gyr has been $<\psi> \sim 3.5 M_\odot pc^{-2} Gyr^{-1}$ and that the ratio $<\psi>/\psi(t_G) \sim 2$. The past SFR is derived simply by subtracting from the solar neighbourhood total surface mass density of $55 M_\odot pc^{-2}$ the amount of gas of $13 M_\odot pc^{-2}$ (see chapter 1) thus obtaining $42 M_\odot pc^{-2}$, which is roughly the mass turned into stars in the last 12 Gyr. Therefore, gas infall would increase the timescales for gas depletion between 16 and 40%. If these numbers are correct, the infall in the solar neighbourhood should be a minor but not negligible effect at the present time.

Unfortunately, the exact law for the gas accretion onto the Galaxy is not known and, in principle, it should be deduced from a good model of galaxy formation. The most simple scenario is to assume that the Galaxy formed by accretion of gas on a free-fall time, but we know that this is a too simplistic picture especially to explain the formation of the disk (see chapter 6).

A more realistic representation of the infall rate is given by dynamical models for the formation of the Galaxy. Larson, in his pioneering work, computed the formation of disk galaxies by means of hydrodynamical calculations including rotation and axial symmetry. He found that the formation of spheroidal components requires shorter timescales than the formation of disks. The main parameters in his calculations are the collapse and the star formation rate, this latter being linked to the previous through the gas density. He pointed out that a fast star formation rate is required to form the spheroidal components whereas a much slower star formation rate is necessary in the disk to allow the gas to settle to a disk before forming stars. In particular, there should be a phase before the formation of the thin disk during which star formation is inhibited. As we have seen in chapter 1, data on abundance ratios in solar neighbourhood stars (i.e. [Fe/O] vs. [O/H]) seem to indicate a possible hiatus in the star formation between the end of the thick disk and the beginning of the thin disk phase. Moreover, Larson found that the timescale for the formation of the disk is much longer than the timescale for the formation of the spheroidal components (halo and bulge) and increases with increasing distance from the centre; the inner parts of the disk form first, and the outer parts form progressively later as gas with higher and higher angular momentum settles into the equatorial plane at larger and larger radii.

In models of galactic chemical evolution we are forced to parametrize the infall rate because of the lack of a real dynamical treatment. Here we briefly summarize

the most common parametrizations adopted in the literature for the infall rate
(usually expressed in terms of surface gas density):

- constant in space and time, which is obviously not very realistic,

- exponentially decreasing in time and constant in space,

- exponentially decreasing in time and also variable in space.

In particular, Chiosi (1980) assumed an infall rate of primordial material of
the form:

$$(\frac{d\sigma_{gas}}{dt})_{infall} = a(R_G)e^{-t/\tau} M_\odot pc^{-2} Gyr^{-1} \qquad (4.1)$$

where τ is a parameter indicating the timescale for the accretion of the galactic
disk, and this parametrization was then adopted by many subsequent studies.
The quantity τ is a free parameter since we do not have any specific indication
about the mechanism of formation of the galactic disk, and it can be assumed
to be an increasing function of the galactocentric distance R_G, in order to
account for the dynamical results of Larson. Matteucci and François (1989)
adopted the eq. (4.1) for the infall law in the galactic disk and suggested that
the timescale of the infall τ is a linear function of the galactocentric distance,
in the sense that it increases with the radius thus producing a situation of
biased infall or *inside-out* disk formation. The quantity $a(R_G)$ is obtained
by imposing that eq. (4.1) reproduces the observed exponential mass surface
density profile (see eq. 6.4).

Generally, the chemical composition of the infalling material is assumed to
be primordial although there have been a few models adopting an enriched
infall. It has been shown that the results obtained with enriched infall do
not differ substantially from those obtained with primordial infall as long as
the metallicity of the infall does not exceed a critical value ($\sim 0.4Z_\odot$, Tosi,
1988). Enriched infall has often been claimed to explain the high astration
factor required for deuterium, if the observations suggesting a high primordial
D abundance have to be trusted. In fact, an infall of gas enriched in heavy
elements would be depleted in deuterium and a concentration of D lower than
the primordial one in the accreting gas helps in diluting the abundance of D
in the ISM.

- Chiappini et al. (1997) introduced the concept of double infall to explain
the evolution of the halo thick-disk on one side and that of the thin-disk on
the other. In particular, they assumed that the halo-thick disk formed first
by means of a relatively fast infall episode whereas the thin disk formed by
means of a completely independent subsequent infall episode occurring on

much longer timescales. In this way, the galactic disk is almost completely formed out of extragalactic primordial gas (see also chapter 6).

The functional form of the infall rate they proposed is:

$$\left(\frac{d\sigma_{gas}}{dt}\right)_{halothick-disk} \propto e^{-t/\tau_1} \tag{4.2}$$

$$\left(\frac{d\sigma_{gas}}{dt}\right)_{thin-disk} \propto e^{-(t-t_{max})/\tau_2} \tag{4.3}$$

where τ_1 and τ_2 are the timescale for the formation of halo-thick disk and thin disk, respectively, and t_{max} represents the time of maximum gas accretion onto the galactic disk.

In Figure 4.1 is shown the behaviour of the double infall rate in the solar neighbourhood for $\tau_1 = 2$ Gyr and $\tau_2 = 8$ Gyr.

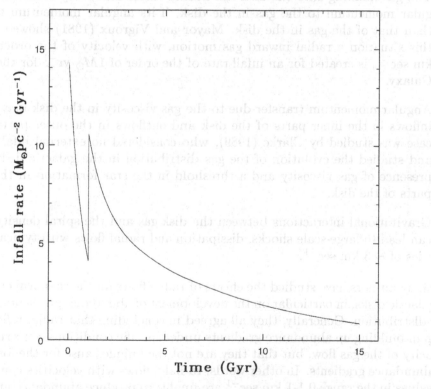

Figure 4.1. The rate of infall of gas (in units of $M_\odot pc^{-2} Gyr^{-1}$) as a function of time predicted by the two-infall model of Chiappini et al. (1997). The first infall episode is responsible for the formation of the halo thick-disk, whereas the second infall episode is responsible for the formation of the thin-disk.

Other authors such as Lacey and Fall (1985) adopted an infall rate exponentially decreasing in time and radius:

$$f(r,t) = \frac{\alpha_f^2 M_D(t_G)}{2\pi} \frac{e^{(-\alpha_f r - t/\tau)}}{\tau[1 - e^{(-t_G/\tau)}]} \tag{4.4}$$

where $M_D(t_G)$ is the total mass of the disk at the present time t_G and α_f and τ are two adjustable parameters. The exponential radial dependence is chosen in order to obtain exponential surface density profiles.

4.2 RADIAL FLOWS ALONG THE GALACTIC DISK

From the physical point of view, radial flows should be expected on the basis of the following arguments:

- The gas infalling onto the disk can induce radial inflows by transferring angular momentum to the gas in the disk, if its angular momentum is lower than that of the gas in the disk. Mayor and Vigroux (1981) showed that in this situation a radial inward gas motion, with velocity of the order of 1-5 km sec^{-1}, is created for an infall rate of the order of $1 M_\odot$ yr^{-1} for the whole Galaxy.

- Angular momentum transfer due to the gas viscosity in the disk may induce inflows in the inner parts of the disk and outflows in the outer parts. This case was studied by Clarke (1989), who considered no external infall of gas and studied the evolution of the gas distribution in the galactic disk in the presence of gas viscosity and a threshold in the star formation in the outer parts of the disk.

- Gravitational interactions between the disk gas and the spiral density waves can lead to large-scale shocks, dissipation and radial flows with typical velocities of ~ 3 km sec^{-1}.

Many authors have studied the effects of radial flows on the chemical evolution of galactic disks, in particular on the development of abundance gradients and the gas distribution. Generally, they all agreed in concluding that radial inflows can help in building up abundance gradients under specific conditions concerning the velocity of the gas flow, but that they are not the unique cause for the formation of abundance gradients. In other words, radial inflows with velocities compatible to values in the range 0.1-1 km sec^{-1} are unable to produce abundance gradients in agreement with the observed ones, in absence of other physical processes. On the other hand, radial outflows generally tend to destroy the gradients (see also chapter 5 and chapter 6).

Lacey and Fall (1985) for example, concluded that radial inflows with velocities not larger than 2 km sec^{-1} contribute to the creation of negative abundance
gradients but only if the dependence of the star formation rate on the gas density is linear. Clarke (1989) concluded that, in absence of external infall, low
velocity radial flows induced by gas viscosity and coupled with a threshold in the
star formation at large galactocentric distances (18 kpc) can well reproduce the
abundance gradient and the gas distribution along the galactic disk. Edmunds
and Greenhow (1995) (see chapter 5) showed that there is no simple "one-way"
effect of flows on gradients, although in the linear star formation case accelerating
inflows tend to flatten the gradients while decelerating inflows tend to steepen
them.

Observations of gas along the galactic disk seem to be consistent with radial
flows of modest velocities up to few km sec^{-1}; the local HI appears to be at rest
relative to the Local Standard of Rest (LSR) to within about 1 km sec^{-1} (Kerr
1969; Crovisier 1978). Concerning the molecular clouds within a few kpc of the
Sun a mean inward motion of roughly 4 km sec^{-1} is found (Stark 1984).

4.3   GALACTIC WINDS

Galactic winds (namely gas outflows from galaxies) are likely to play a fundamental role in the evolution of galaxies as their existence is indicated by several
observational constraints. Galactic winds are now observed in dwarf irregular
starburst galaxies and the chemical composition of the ICM (measured from X-
ray emission lines) shows an almost solar iron abundance indicating that the
galaxies in clusters should loose a substantial amount of their ISM.

The main cause for galactic winds are probably the supernovae which can
trigger gas outflows when the thermal content of the gas, resulting from energy
deposition from their explosions, equals the binding energy of the gas. This
is a necessary condition for the occurrence of the wind but not sufficient, in
fact, for a wind to really develop the energy deposited into the ISM should not
be radiated away. To this purpose, a good condition is that all the supernova
remnants should overlap, in other words that the filling factor should be equal
to the unity. In this situation, in fact, the SNe of the second generation would
explode in a rarified and hot medium and transfer the maximum energy since
the radiation is proportional to the square of the gas density.

Stellar winds from massive stars also contribute to the heating of the ISM,
although their contribution, in most of the astrophysical situations, is substantially smaller than that of supernovae, with the exception of the very early stages
of a starburst, as shown by Leitherer et al. (1992).

From a theoretical point of view, galactic winds have been originally suggested
by Mathews and Baker (1971) and then by Larson (1974) to explain the observed
·mass-metallicity relation in elliptical galaxies and the amount of iron in the ICM
(Matteucci and Vettolani 1988 and references therein) as well as to reproduce the

observed scatter in the properties of dwarf irregular and blue compact galaxies (Marconi et al. 1994 and references therein). It is not yet clear whether galactic winds can be important in the evolution of galactic disks. This does not seem to be the case, at least for our Galaxy, as we will see in chapter 6. In fact, models of chemical evolution of the galactic disk with mass outflows do not seem to reproduce the abundances and abundance ratios observed in field stars (Tosi et al. 1998).

Supernova-driven galactic wind models have become very popular in the last ten years since they seem to reproduce well the chemical and photometric properties of elliptical galaxies (Gibson, 1997 and references therein) together with the observed abundances of heavy elements in the ICM. The ordinary galactic wind is made of ambient ISM where the metals produced and ejected by stars are normally well mixed with the pre-existing medium. Therefore, in general, the element abundances in the wind should reflect those of the ISM. However, supernova explosions can produce chimneys which eject outside the galaxy mostly the SN ejecta, namely metal enriched material. This possibility has been explored by several theoretical papers showing that in dwarf starburst galaxies this is a very likely occurrence (De Young and Gallagher, 1990: Mac Low and Ferrara, 1999; D'Ercole and Brighenti 1999).

However, these interesting results should still be regarded as preliminary since we know very little about the interactions between the evolving supernova remnants, massive star bubbles and the ISM. In particular, it is very difficult to assess how much energy is deposited from winds and supernovae into the ISM and how much is lost. This is a key problem in the field of galaxy formation and evolution that will be certainly developed in the future thanks to the advent of more and more sophisticated dynamical and chemical evolution models. Another fundamental parameter which enters in the development of a galactic wind is the potential well of the considered galaxy and therefore involves the problem of dark matter halos.

At the moment we can only parametrize the galactic winds and here we recall the most common parametrizations :

- *a)* a continuous wind rate proportional to the star formation rate, a formulation which has been applied in models predicting the evolution of starburst galaxies, namely:

$$(\frac{d\sigma_{gas}}{dt})_{wind} = w\psi(t) \qquad (4.5)$$

where w is a free parameter measuring the *efficiency* of the galactic wind. If we want to express the rate of galactic wind per chemical element, i, we can write:

$$\left(\frac{d\sigma_{gas}X_i}{dt}\right)_{wind} = w_i\psi(t) \tag{4.6}$$

where X_i is the ISM abundance of the element i. The efficiency w_i is assumed to be constant.

- *b)* A continuous wind rate proportional to the star formation rate but where only certain elements are lost, or where different elements are lost in different proportions. In order to do that one can simply assume w_i to be different for different chemical elements, for example to be zero for the elements produced mostly by single low and intermediate mass stars, such as N and He, and to be different from zero for the α-elements which are mainly produced in type II SNe (Marconi et al. 1994). In real situations we would expect that both types of wind, namely normal and enriched, are present (Pilyugin 1993).

Enriched galactic winds can explain some anomalous abundance ratios observed in some starburst galaxies, such as for example the dwarf galaxy IZw18, where is observed an almost solar N/O ratio, very difficult to reproduce by adopting the standard stellar nucleosynthesis and normal galactic winds. Another example is the high $\Delta Y/\Delta Z$ ratio (namely the enrichment of helium relative to metals during the galactic lifetime) observed in dwarf irregular and blue compact galaxies. In this case, an enriched wind, of the type described before, produces a good agreement with the observed $\Delta Y/\Delta Z$.

- *c)* A sudden wind occurring when the thermal energy of the gas exceeds its binding energy and all the gas present is lost. This is the case of the supernova-driven models adopted for studying the evolution of elliptical galaxies (Arimoto and Yoshii, 1987; Matteucci and Tornambè, 1987).

Chapter 5

BASIC EQUATIONS OF CHEMICAL EVOLUTION

5.1 ANALYTICAL MODELS

First of all, we discuss the so-called *Simple Model* for the chemical evolution of the solar neighbourhood. We note that the solar neighbourhood is defined as a region centered in the Sun and extending roughly 1 kpc in all directions.

5.1.1 BASIC ASSUMPTIONS OF THE SIMPLE MODEL

In this section we follow the definition of the *Simple Model* as given in Tinsley (1980); in particular, the *Simple Model* is based on the following assumptions:

- 1 the system is one-zone and closed, namely there are no inflows or outflows
- 2 the initial gas is primordial
- 3 $\varphi(m)$ is constant in time
- 4 the gas is well mixed at any time.

Following the formalism of Tinsley (1980) we define:

$$\mu = \frac{M_{gas}}{M_{tot}} \tag{5.1}$$

as the fractional mass of gas, with:

$$M_{tot} = M_* + M_{gas} \tag{5.2}$$

where M_* is the mass in stars (dead and living). Possible non-baryonic dark matter is not considered.

From eq. (5.1) and (5.2) it follows that:

145

$$M_* = (1 - \mu)M_{tot} \tag{5.3}$$

The metallicity is defined by:

$$Z = \frac{M_Z}{M_{gas}} \tag{5.4}$$

where M_Z is the mass in the form of metals (all the elements heavier than He). The initial conditions are:

$$M_{gas}(0) = M_{tot} \tag{5.5}$$

$$Z(0) = 0 \tag{5.6}$$

The equation for the evolution of the gas in the system is:

$$\frac{dM_{gas}}{dt} = -\psi(t) + E(t) \tag{5.7}$$

where $E(t)$ is the rate at which dying stars restore both the enriched and unenriched material into the ISM at the time t. This quantity can be written as:

$$E(t) = \int_{m(t)}^{\infty} (m - M_R)\psi(t - \tau_m)\varphi(m)dm \tag{5.8}$$

where $m(t)$ is the mass born at the time $t = 0$ and dying at the time t, $m - M_R$ is the total mass ejected from a star of initial mass m and τ_m is the lifetime of a star of initial mass m. When eq. (5.8) is substituted in eq. (5.7) one has an integer-differential equation which can be solved analytically only by making a simplifying assumption, namely by assuming I.R.A. as already defined in Section 2.8. This approximation allows us to neglect the stellar lifetimes in eq. (5.8) and to define the following quantities:

$$R = \int_{1}^{\infty} (m - M_R)\varphi(m)dm \tag{5.9}$$

which is the total mass fraction (we speak of fraction instead of total mass because of the normalization adopted for the IMF, see chapter 3) which is restored to the ISM by a stellar generation. This quantity is a constant unless one assumes that the amount of mass lost by a star is a strong function of metallicity as is the case for the massive star yields of M92. Under rather standard assumptions concerning the IMF and the $m - M_R$ relation one can find that $R = 0.20 - 0.50$. The other important quantity is the *galactic yield* as defined already in Section 2.8:

$$y_Z = \frac{1}{1-R} \int_1^\infty m p_{Zm} \varphi(m) dm, \tag{5.10}$$

where the quantity p_{Zm} is the mass fraction of the newly produced and ejected metals by a star of mass m.

Equation (5.8) can now be written as :

$$E(t) = \psi(t)R \tag{5.11}$$

and consequently eq. (5.7) as:

$$\frac{dM_{gas}}{dt} = -\psi(t)(1-R) \tag{5.12}$$

Now we write the equation for the evolution of metals:

$$\frac{d(ZM_{gas})}{dt} = -Z\psi(t) + E_Z(t) \tag{5.13}$$

where:

$$E_Z(t) = \int_{m(t)}^\infty [(m-M_R)Z(t-\tau_m) + m p_{Zm}]\psi(t-\tau_m)\varphi(m)dm \tag{5.14}$$

where the first term in the square brackets represents the mass of pristine metals which are restored into the ISM without suffering any nuclear processing, whereas the second term contains the newly formed and ejected metals. Under the assumption of I.R.A. equation (5.14) becomes:

$$E_Z(t) = \psi(t)RZ(t) + y_Z(1-R)\psi(t) \tag{5.15}$$

Eq. (5.15) is then substituted in eq. (5.13) which is solved analytically.

It can be easily demonstrated that equation (5.13) has the following solution:

$$Z = y_Z ln(\frac{1}{\mu}) \tag{5.16}$$

obtained after integrating between $M_{gas}(0) = M_{tot}$ and $M_{gas}(t)$ and $Z(0) = 0$ and $Z(t)$.

The yield which appears in eq. (5.16) is known as *effective yield*, simply defined as the yield $y_{Z_{eff}}$ that would be deduced if the system were assumed to be described by the Simple Model.

Therefore:

$$y_{Z_{eff}} = \frac{Z}{ln(1/\mu)} \tag{5.17}$$

If $y_{Z_{eff}} > y_Z$ then the actual system has attained a higher metallicity for a given gas fraction μ.

5.1.2 THE FRACTION OF ALL STARS WITH METALLICITIES UP TO Z

From equation (5.3) we can calculate the fraction of all stars that had been formed while the gas fraction was $\geq \mu$:

$$\frac{M_*}{M_1} = \frac{1-\mu}{1-\mu_1} \tag{5.18}$$

where the subscript 1 represents the present time value. Having in mind equation (5.16) we see that all of these stars were formed with metallicities $\leq y_Z ln\mu^{-1}$. Therefore, the fraction of all stars with metallicities $\leq Z$, indicated by $S(Z)$, can be written as:

$$S(Z) = \frac{1 - e^{-Z/y_Z}}{1 - \mu_1} \tag{5.19}$$

It is useful to eliminate the yield y_Z from this expression by using the relation $y_Z = \frac{Z_1}{ln\mu_1^{-1}}$ so that $S(Z)$ can be written in terms of the ratio Z/Z_1 and the present time gas fraction:

$$S(Z) = \frac{1 - \mu_1^{Z/Z_1}}{1 - \mu_1} \tag{5.20}$$

If one compares this distribution with a log-normal approximation to the data for stars in the solar neighbourhood, one finds that the Simple Model of chemical evolution predicts too many metal poor stars than observed. This is known as the "G-dwarf problem" (see Figure 5.1) and shows how one or more assumptions of the Simple Model need to be relaxed. One of the most obvious solutions to the *G-dwarf problem* is to relax the assumption that the solar neighbourhood evolves as a closed box and allow for the existence of gas flows. In particular, by assuming that the solar vicinity is formed by slow infall of gas one can obtain a good agreement with the observational data. This kind of model has been extensively studied and it will be discussed later in great detail. In Figure 5.1 is also shown the case called *extreme infall*, namely when the infall rate exactly balances the SFR plus the rate of matter restitution from dying stars.

5.1.3 AVERAGE STELLAR METALLICITY

Generally, in the chemical evolution of galaxies the metallicity, as defined before, refers to the metal content of the interstellar medium. However, in some cases, it is useful and necessary to know the average metallicity of all the stars in a galaxy, namely the average stellar metallicity. This is defined as:

$$< Z >_M = \frac{1}{S_1} \int_0^S Z(S')dS' \tag{5.21}$$

where S_1 is the fraction of all stars with metallicity $\leq Z_1$, namely the present time metallicity. It is worth noting that this is a mass averaged metallicity. In reality one measures the luminosity averaged metallicity in galaxies, for example from integrated spectra of elliptical galaxies. Therefore, it is useful to define the luminosity averaged metallicity:

$$< Z >_L = \frac{\sum_{i,j} n_{ij} Z_i L_{V_j}}{\sum_{i,j} n_{ij} L_{V_j}} \tag{5.22}$$

where n_{ij} represents the number of stars in the metallicity interval Z_i and in the visual luminosity interval L_{V_j}. In fact, the metallicity indices measured from integrated spectra represent the metallicity of the stellar population which predominates in the visual light. In order to obtain $< Z >_L$, one needs a photometric model to calculate the stellar luminosities. galactic evolution models (Arimoto and Yoshii, 1987; Matteucci and Tornambè, 1987; Greggio, 1997) have shown that the two average metallicities are different, since metal poor giants tend to predominate in the visual luminosity so that, in general, the luminosity-averaged metallicity is smaller than the mass-averaged one. However, the differences between the two averages tend to decrease as the galactic mass increases, owing to the presence of fewer and fewer metal poor giants.

5.1.4 SECONDARY ELEMENTS

We remind here that a chemical element is *secondary* if it is produced proportionally to the initial metallicity in the star. Typical secondary elements are ^{14}N and ^{13}C which are the by-products of the CNO-cycle and the *s-process* elements which are produced by neutron capture on the pre-existing Fe nuclei. Actually, they are *tertiary* elements since the neutron flux depends itself upon the stellar metallicity Z.

Let us consider the abundance of a secondary element X_S with a primary seed with abundance Z, and be p_{sm} the fraction of matter produced and restored into the ISM by a star of mass m and metallicity $Z_\odot$ in the form of the secondary element. For a generic metallicity Z the yield of the secondary element produced and ejected by a star of mass m is $p_{sm}(Z/Z_\odot)$. Therefore, the matter ejected in the form of the secondary element by an entire stellar generation is:

$$E_S = \int_{m(t)}^{\infty} [(m - M_R) X_S(t - \tau_m) + m p_{sm} \frac{Z(t - \tau_m)}{Z_\odot}] \psi(t - \tau_m) \varphi(m) dm \tag{5.23}$$

By substituting eq. (5.23) into (5.13) and assuming the I.R.A. one obtains:

$$M_{gas} \frac{dX_S}{dt} = y_S (Z/Z_\odot)(1 - R)\psi \tag{5.24}$$

which transforms into:

$$M_{gas}\frac{dX_S}{dM_{gas}} = -y_S(\frac{Z}{Z_\odot}) \tag{5.25}$$

after dividing by (5.12). This equation can be integrated analytically under the initial conditions of the simple closed-box model to give:

$$X_S = y_S(\frac{Z}{Z_\odot})ln\mu^{-1} \tag{5.26}$$

which becomes:

$$X_S = \frac{1}{2}(\frac{y_S}{y_Z Z_\odot})Z^2 \tag{5.27}$$

having used the equation (5.16). This means that for a secondary element the simple closed-box model predicts that its abundance increases proportionally to the metallicity squared, namely:

$$X_S \propto Z^2 \tag{5.28}$$

Consideration of the infall changes equation (5.28), in particular in the case of the extreme infall the abundance of a secondary element grows even more rapidly:

$$X_S = (\frac{y_Z y_S}{Z_\odot})(1 - e^{-\nu} - \nu e^{-\nu}) \tag{5.29}$$

where $\nu = \mu^{-1} - 1$.

From the discussion above we infer that, if we apply to a galaxy or to a galactic region the solutions of the simple closed box model with I.R.A., we should expect a primary element with abundance $[X_P/H]$ to evolve independently of the initial stellar metallicity $[Z/H]$, namely $[X_P/Z] = cost$ over the whole range of $[Z/H]$, whereas we expect a secondary element to behave in such a way that the ratio of its abundance relative to $[Z/H]$ grows linearly with $[Z/H]$, namely $[X_S/Z] \propto [Z/H]$. This is clearly an oversimplification, first of all because ignoring the stellar lifetimes is a poor approximation for elements mostly produced by long-lived stars as shown by the observed abundances of secondary elements which do not behave as predicted by the simple model. In fact, if we take the ratio between a typical secondary element and a typical primary element such as $[Ba/Fe]$, we should expect on the basis of equation (5.27) that $[Ba/Fe] \propto [Fe/H]$ whereas the observations show that $[Ba/Fe]$ is constant and solar in the range $-2.0 \leq [Fe/H] \leq 0$ and it declines steadily for $[Fe/H] < -2.0$ (see Figure 1.5).

5.1.5 EQUATIONS CONTAINING GAS FLOWS

The equation for the evolution of metals in presence of gas flows can be written as:

$$\frac{d(ZM_{gas})}{dt} = -Z(t)\psi(t) + E_Z(t) + Z_A A(t) - Z(t)W(t) \qquad (5.30)$$

where $A(t)$ is the accretion rate of matter with metallicity Z_A and $W(t)$ is the rate of loss of material from the system. If $A(t) = W(t) = 0$ equation (5.30) becomes equation (5.13).

Let us examine now the case of only outflow:

a) $A(t) = 0$, $W(t) \neq 0$.

The fundamental equations in this case are:

$$\frac{dM_{tot}}{dt} = -W(t) \qquad (5.31)$$

$$\frac{dM_{gas}}{dt} = -(1-R)\psi(t) - W(t) \qquad (5.32)$$

$$\frac{d(ZM_{gas})}{dt} = -(1-R)\psi(t)Z(t) + y_Z(1-R)\psi(t) - Z(t)W(t) \qquad (5.33)$$

The easiest way of defining $W(t)$ is to assume that it is proportional, through a constant λ, to the rate of star formation plus the rate of recycling of material from stars:

$$W(t) = \lambda(1-R)\psi(t) \qquad (5.34)$$

where $\lambda \geq 0$ is the wind parameter. In analogy with the case with no flows, after some algebraic manipulations one arrives at:

$$\frac{dZ}{dM_{gas}} = -\frac{y_Z}{M_{gas}(1+\lambda)} \qquad (5.35)$$

which can be integrated between 0 e $Z(t)$ and between $M_{tot} = M_{gas}(0)$ and $M_{gas}(t)$ obtaining:

$$Z = \frac{y_Z}{(1+\lambda)}ln[(1+\lambda)\mu^{-1} - \lambda] \qquad (5.36)$$

It is clear that for $\lambda = 0$ eq. (5.36) becomes eq. (5.16). The meaning of eq. (5.36) is immediately clear, the true yield is always lower than the effective yield in presence of outflows.

The opposite case:

b) $A(t) \neq 0$ and $W(t) = 0$ leads to the following equations:

$$\frac{dM_{tot}}{dt} = A(t) \qquad (5.37)$$

$$\frac{dM_{gas}}{dt} = -(1 - R)\psi(t) + A(t) \tag{5.38}$$

$$\frac{d(ZM_{gas})}{dt} = -(1 - R)\psi(t)Z(t) + y_Z(1 - R)\psi(t) + Z_A A(t) \tag{5.39}$$

where the accretion rate has been chosen to be:

$$A(t) = \Lambda(1 - R)\psi(t) \tag{5.40}$$

and Λ is a positive constant different from zero. After some algebraic manipulations one obtains:

$$M_{gas}\frac{dZ}{dt} = y_Z(1 - R)\psi(t) + (Z_A - Z)A(t) \tag{5.41}$$

The solution of the eq. (5.41) for a primordial infalling material ($Z_A = 0$) and $\Lambda \neq 1$ is:

$$Z = \frac{y_Z}{\Lambda}[1 - (\Lambda - (\Lambda - 1)\mu^{-1})^{-\Lambda/(1-\Lambda)}] \tag{5.42}$$

If $\Lambda = 1$ the solution is: $Z = y_Z[1 - e^{-(\mu^{-1}-1)}]$, which is the well-known solution for the *extreme infall* (Larson, 1972), where the amount of gas remains constant in time. The quantity $\mu^{-1} - 1$ represents the ratio between the accreted mass and the initial mass. The extreme-infall solution shows that when $\mu \to 0$ $Z \to y_Z$.

In Figure 5.1 one can see that the extreme infall case is not a good solution for the G-dwarf problem since it predicts too few metal poor stars when compared with the observational data.

Finally, in the case in which both infall and outflow are present, namely the solution of the complete equation (5.30) is:

$$Z = \frac{y_Z}{\Lambda}\{1 - [(\Lambda - \lambda) - (\Lambda - \lambda - 1)\mu^{-1}]^{\frac{\Lambda}{\Lambda-\lambda-1}}\}, \tag{5.43}$$

for a primordial infalling gas ($Z_A = 0$).

5.1.6 TIMESCALES OF CHEMICAL EVOLUTION

It is interesting to define some particular timescales which are important for the chemical evolution of galaxies.

The first timescale we define is the *star formation timescale*, namely the time necessary to consume all the gas in a given system. This time is defined as:

$$\tau_* = \frac{M_{gas}}{|dM_{gas}/dt|} = \frac{M_{gas}}{(1 - R)\psi} \tag{5.44}$$

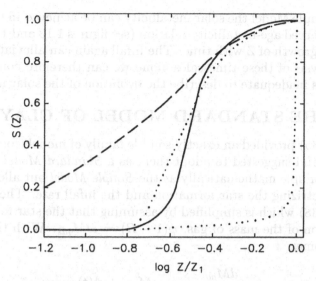

Figure 5.1. Cumulative stellar metallicity distribution. S(Z) is the fraction of stars having metallicities $\leq Z$, with a maximum value Z_1 (taken as $2Z_\odot$). *Solid line*: log-normal representation of the data for the solar neighbourhood. *Long dashes*: the prediction of the Simple Model for the chemical evolution. *Short dashes*: the prediction of the extreme infall model. *Dots*: prediction of a model with a finite initial metallicity. From Tinsley 1980, Fund. Cosmic Phys. Vol. 5, 287; reproduced here by kind permission of Gordon & Breach Science Publishers (copy right 1980).

under the assumption that the Simple Model holds. If we assume the surface mass density in the solar neighbourhood to be $\simeq 10 M_\odot pc^{-2}$, the present time star formation rate to be $\simeq 10 M_\odot pc^{-2} Gyr^{-1}$ and $(1 - R) = 0.80$, we obtain $\tau_{*SN} \simeq 1$ Gyr. This means that if the solar neighbourhood evolves as the Simple Model predicts, we expect the gas to be consumed in a billion year from now. Therefore, we should conclude that we are in the final phase of the evolution of the solar vicinity, which is highly unlikely. On the basis of this consideration the existence of infall onto the galactic disk has been claimed. In fact, if we adopt an infall model instead of the simple model the value of τ_* is longer.

Another interesting timescale is the *timescale for the chemical enrichment*, namely the time necessary to achieve a precise metallicity in the gas:

$$\tau_Z = \frac{Z}{|dZ/dt|} = \frac{ZM_{gas}}{|y_Z(1 - R)\psi|} \tag{5.45}$$

which can be written as:

$$\tau_Z = \tau_*(Z/y_Z) \simeq \tau_* \simeq 10^9 \ yr \tag{5.46}$$

and assuming that the metallicity Z and the yield y_Z are comparable at the present time in the solar neighbourhood. This result implies that in the framework of the Simple Model the solar metallicity can be achieved in only one billion years. The observed age-metallicity relation (see figures 1.13 and 1.14) implies instead a slower growth of Z with time. The infall again can alleviate the problem. From the analysis of these timescales alone we can therefore conclude that the simple model is inadequate to describe the evolution of the solar neighbourhood.

5.1.7   THE STANDARD MODEL OF CLAYTON

Clayton (1985) provided an exactly soluble family of models for galactic chemical evolution and suggested to adopt them as a *Standard Model*. These models are almost as simple mathematically as the *Simple Model* but allow a large freedom in parametrizing the star formation and the infall rate. The starting point is equation (5.38) which is simplified by assuming that the star formation rate is a linear function of the mass of gas, $\psi(1 - R) = \omega M_{gas}$. With this assumption eq. (5.38) becomes:

$$\frac{dM_{gas}}{dt} = -\omega M_{gas} + A(t) \tag{5.47}$$

where ω is a constant.

The solution of eq. (5.47) is:

$$M_{gas}(t) = M_{gas}(0)e^{-\omega t + \int_0^t \frac{A(t')}{M_{gas}(t')}dt'}, \tag{5.48}$$

where $M_{gas}(0)$ is the initial mass of gas in the system. The ratio $\frac{A(t)}{M_{gas}(t)} = \omega_A(t)$ is the instantaneous rate at which the current $M_{gas}(t)$ is being replenished by the current infall $A(t)$. If one defines a dimensionless function of time:

$$\theta(t) = \int_0^t \omega_A(t')dt', \tag{5.49}$$

where $\theta = \overline{\omega_A}t$ represents the number of times that the ISM has been cycled by infall, then eq. (5.48) becomes:

$$M_{gas}(t) = M_{gas}(0)e^{-\omega t + \theta(t)} \tag{5.50}$$

From eq. (5.50) it is clear that M_{gas} first increases if $\theta(t)$ initially increases faster than ωt.

The mass of stars $M_*(t)$ is given by:

$$M_*(t) = \omega M_{gas}(0) \int_0^t e^{-\omega t' + \theta(t')}dt' \tag{5.51}$$

In this framework the equation for metals (eq. 5.39) becomes:

$$\frac{dZ}{dt} = y_Z\omega - (Z - Z_A)d\theta/dt \tag{5.52}$$

the solution of which is:

$$(Z - Z_A) = e^{-\theta}[y_Z\omega \int_0^t e^{\theta(t')}dt' + Z_0 - Z_A] \tag{5.53}$$

where Z_0 is the initial abundance of gas at t=0. Generally, it is reasonably assumed that $Z_0 = 0$ and $Z_A = 0$.

Clayton (1985, 1987) studied the properties of the solutions of the standard model in the case that $A(t)$ has the specific form:

$$A(t) = \frac{k}{t + \Delta}M_{gas}(t) = \frac{kM_{gas}(0)}{\Delta}(\frac{t + \Delta}{\Delta})^{k-1}e^{-\omega t} \tag{5.54}$$

where k and Δ are two free parameters which allow adequate form-fitting to ω_A. With this form for the infall rate, eq. (5.53) becomes:

$$(Z - Z_A) = \frac{y_Z\omega\Delta}{k+1}[(\frac{t + \Delta}{\Delta}) - (\frac{t + \Delta}{\Delta})^{-k}] + (\frac{\Delta}{t + \Delta})^k(Z_0 - Z_A) \tag{5.55}$$

If $Z_0 = 0$ and $Z_A = 0$ then:

$$Z = \frac{y_Z\omega\Delta}{k+1}[(\frac{t + \Delta}{\Delta}) - (\frac{t + \Delta}{\Delta})^{-k}] \tag{5.56}$$

This *Standard Model* is useful since it allows one to study analytically the model response to different histories of the star formation and infall rate, representing a step forward relative to the analytical solutions described before, which do not take into account explicitly the star formation rate and require a very specific assumption for the rate of gas flows. In general, the described analytical models are useful to understand the physical behaviour of gas in galactic systems without appealing to numerical models.

An interesting application of Clayton's model is due to Pagel (1989) who introduced a fixed time delay to study the evolution of the abundances of elements produced on long timescales, such as Fe, in the framework of I.R.A. In particular, for the linear star formation rate, discussed above, coupled with a fixed time delay Δ he found:

$$Z(t) = y_Z e^{(1+\lambda)\omega\Delta}\omega(t - \Delta) \tag{5.57}$$

for $t > \Delta$, where y_Z is the true yield and λ is the wind parameter as defined in eq. (5.40); this parameter takes into account the formation of the galactic disk out of gas lost from the halo. The equation (5.57) is a better approximation than the model with I.R.A. for describing the evolution of the Fe abundance, but is still a rough approximation if compared with numerical models including the detailed treatment of the type Ia SN rate where the different stellar lifetimes are taken into account (see next sections).

5.1.8   EQUATIONS WITH RADIAL FLOWS

In general, radial flows can occur in the disks of spirals as a result of angular momentum transfer by means of viscosity or gravitational interaction with a density wave in the stellar disk. This process tends to concentrate the metals towards the center of the galactic disk and could, in principle, be important for the formation of abundance gradients along the disks.

We assume the formalism of Tinsley (1980) and consider a ring in the galactic disk between the radius r and $r + \delta r$. The chemical evolution of this ring can be studied by means of the equations described above. In particular, let us substitute $M_{gas} \to 2\pi r M_{gas} \delta r$ and $\psi \to 2\pi r \psi \delta r$, where M_{gas} and ψ are now expressed as surface densities.

The net radial flow is then:

$$F(r) - F(r + \delta r) = -(\partial F/\partial r)\delta r \tag{5.58}$$

where F is expressed in $M_\odot yr^{-1}$ and it has a positive sign if directed outwards.

In analogy, the metal flow is defined as:

$$Z(r)F(r) - Z(r + \delta r)F(r + \delta r) = -Z(\partial F/\partial r)\delta r - (\partial Z/\partial r)F\delta r \tag{5.59}$$

If we substitute these expressions for the gas and metal flow in the chemical evolution equations, in particular in the eq. (5.38) and (5.39) where the radial flow substitutes the infall term $A(t)$, we obtain:

$$\frac{\partial M_{gas}}{\partial t} = -(1 - R)\psi - \frac{1}{2\pi r}\frac{\partial F}{\partial r} \tag{5.60}$$

and

$$M_{gas}\frac{\partial Z}{\partial t} = y_Z(1 - R)\psi - \frac{1}{2\pi r}\frac{\partial Z}{\partial r}F \tag{5.61}$$

From the previous equation one can see that a radial flow is consistent with an almost steady-state abundance gradient:

$$\frac{\partial(Z/y_Z)}{\partial r} \sim 2\pi r\frac{(1 - R)\psi}{F} \tag{5.62}$$

According to the way we have defined the radial flow, the abundance gradient will be positive if the flow goes outwards and negative in the opposite case. Since in the disks of spirals, including the Milky Way, are observed negative gradients for metals, we are interested only in inwards radial flows. The radial flow will induce the metallicity Z to change on a timescale given by:

$$\tau_F \sim 2\pi r^2\frac{M_{gas}}{|F|} \tag{5.63}$$

We express now F in terms of velocity of the flow, namely:

$$|F| = 2\pi r M_{gas}|v_F| \tag{5.64}$$

where v_F is the velocity of the gas flow. This implies:

$$\tau_F \sim \frac{r}{|v_F|} \tag{5.65}$$

and therefore:

$$\frac{\partial(Z/y_Z)}{\partial \ln r} \sim \frac{\tau_F}{\tau_*} \tag{5.66}$$

We can clearly see that since τ_F is inversely proportional to $|v_F|$, the metallicity gradient will be favored by a slow gas flow. In particular, several chemical evolution studies have suggested that the gradient disappears if the velocity of the flow is $|v_F| > 2$ km sec^{-1}.

5.1.9 EDMUNDS' THEOREMS

In a series of papers Edmunds and collaborators (Edmunds 1990; Edmunds and Greenhow 1995) studied the effects of gas flows (including radial flows) on the chemical evolution of galaxies through analytical models. In these papers, where we address the reader for a detailed description, were stated and proven several useful theorems. The starting point is the definition of effective yield as shown in eq. (5.17) and then on the basis of what discussed above the theorems are quite intuitive.

The theorems are:

- T(1) In a model with outflow but no infall (or inflow), the effective yield is always less than that of the Simple Model.

- T(2) In a model with infall whose metallicity does not exceed that of the system itself, the effective yield is always less than that of the Simple Model provided that the outflow rate exceeds the infall rate.

- T(3) In a model with infall of unenriched gas, the effective yield is always less than that of the Simple Model.

- T(4) The G-dwarf problem cannot be solved by any outflow.

- T(5) The G-dwarf problem can be solved by particular forms of infall.

- T(6) In a model with outflow but no infall, the ratio of secondary to primary elements is unchanged from that expected in the Simple Model.

- T(7) The mass-weighted mean stellar metallicity in a model with outflow is always less than that in the Simple Model.

- T(8) The mass-weighted mean stellar metallicity in a model with unenriched infall is always less than, or (in the limit of gas exhaustion) equal to, the true yield y_Z.

- T(9) For a galaxy in which star formation is proportional to the first power of the gas density, and in which there is no accretion of material, the chemical evolution of the gas with any radial flow (which is not a function of gas density) is independent of the initial or subsequent distribution of gas.

- T(10) For a galaxy with no accretion in which the star formation rate is simply given by a function of t times the first power of the gas density, no abundance gradient is generated in the gas for any radial flow velocity that is not a function of gas density.

- T(11) In the absence of any gas flow or accretion, the gas abundance gradient is always steeper than the stellar abundance gradient.

- T(12) Abundance gradients caused by radial variation of the star formation rate may be locally or globally shallower, steeper or be unaffected by radial flow.

Theorems T(1) and T(2) can be visualized in Figure 5.2 where Z is plotted as a function of $\ln \mu^{-1}$: the straight line represents the predictions of the Simple Model and it gives the upper limit for Z at any μ. Models with infall (or inflow) of unenriched gas and outflow can populate only the region below this line.

5.2   NUMERICAL MODELS

A complete chemical evolution model in the presence of both galactic wind and infall can be described by a number of equations equal to the number of chemical species: in particular, if $G_i = \frac{\sigma_i}{\sigma(t_G)}$ is the mass fraction of the gas in the form of any chemical element i, we can write:

$$\dot{G}_i(t) = -\psi(t)X_i(t) + \int_{M_L}^{M_{Bm}} \psi(t - \tau_m)Q_{mi}(t - \tau_m)\varphi(m)dm + \qquad (5.67)$$

$$A \int_{M_{Bm}}^{M_{BM}} \varphi(m)[\int_{\mu_{min}}^{0.5} f(\mu)\psi(t - \tau_{m2})Q_{mi}(t - \tau_{m2})d\mu]dm +$$

$$(1 - A) \int_{M_{Bm}}^{M_{BM}} \psi(t - \tau_m)Q_{mi}(t - \tau_m)\varphi(m)dm +$$

$$\int_{M_{BM}}^{M_U} \psi(t - \tau_m)Q_{mi}(t - \tau_m)\varphi(m)dm + X_{iA}(t)A(t) - X_i(t)W(t)$$

The quantity $X_i(t) = \sigma_i(t)/\sigma_{gas}(t)$ represents the abundance by mass of the element i and by definition the summation over all the abundances of the elements

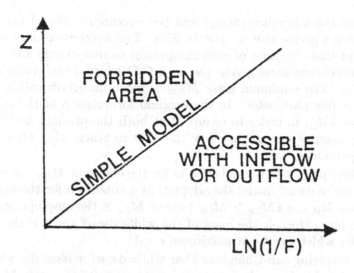

Figure 5.2.   Schematic plot of metallicity Z versus $\ln(1/f)$. Here f is the equivalent of $\mu = \frac{M_{gas}}{M_{tot}}$. The area which can be reached by unenriched infall or outflow is shown. From Edmunds 1990, M.N.R.A.S. Vol. 246, 678; reproduced here by kind permission of Blackwell Science LTD (copy right 1990).

present in the gas mixture is equal to unity. We define the total surface mass density of gas, $\sigma(t) = \sigma_{gas} + \sigma_*$, as the mass of stars (dead and living) plus gas, namely $\sigma(t_G)$ is the present time total surface mass density. The quantity $W_i(t) = X_i(t)W(t)$ represents the rate at which the element i is lost through a galactic wind, whereas the quantity $A_i = X_{iA}(t)A(t)$ is the rate at which the element i is accreted through infall. Generally one assumes that the abundances of the infalling gas, $X_{iA}(t)$, are constant in time. The SFR is $\psi(t)$, which should be expressed as a function of $G = \frac{\sigma_{gas}}{\sigma(t_G)}$ (the mass fraction of the gas), and the IMF is $\varphi(m)$. The rates $A(t)$ and $W(t)$ should also be expressed as functions of the mass fraction of the gas. In other words, every term of eqs. (5.67) is normalised to $\sigma(t_G)$.

The various integrals appearing in eq. (5.67) represent the rate of restitution of matter from the stars into the ISM and they include the contribution from type Ia SNe as introduced by Matteucci and Greggio (1986).

In particular, the second integral refers to the material restored by type Ia SNe, in other words by all the binary systems which have the right properties to give rise to a type Ia SN event. The rate is calculated by assuming the single-degenerate model (a C-O WD plus a red giant companion) for the progenitors of these SNe with the prescriptions of Greggio and Renzini (1983a). This kind of formulation is now widely adopted and therefore it is worth mentioning its basic assumptions. The extremes of the second integral on the right side of eq.

(5.67) represent the minimum (M_{Bm}) and the maximum (M_{BM}) mass allowed for binary systems giving rise to type Ia SNe. The maximum mass is fixed by the requirement that the mass of each component cannot exceed $8M_\odot$, which is the assumed maximum mass giving rise to a C-O white dwarf, thus leading to $M_{BM} = 16M_\odot$. The minimum mass M_{Bm} is more uncertain and it has been considered as a free parameter. In the original formulation of the type Ia SN rate is $M_{Bm} = 3M_\odot$, in order to ensure that both the primary and secondary star would be massive enough to allow the WD to reach M_{Ch} after accretion from the companion.

Greggio (1996) presented revised criteria for the choice of M_{Bm}: in particular, for the single white dwarf model she adopted as a condition for the explosion of the system that $M_{WD} + \epsilon M_{2,e} \geq M_{Ch}$ (where $M_{2,e}$ is the envelope mass of the evolving secondary, M_{WD} is the mass of the white dwarf and ϵ is the efficiency of mass transfer which can be at maximum $\epsilon = 1$).

For models involving sub-Chandrasekhar white dwarf masses she adopted the following criterion for the explosion to occur: $M_{WD} \geq 0.6M_\odot$ and $\epsilon M_{2,e} \geq 0.15$. She also showed that in order to ensure having SN explosions at late times in the framework of an initial burst of star formation, as required in ellipticals, the value of ϵ has to be larger than 0.5 for Chandrasekhar white dwarfs whereas an efficiency as low as 0.25 is enough for sub-Chandrasekhar models.

The function $f(\mu)$ describes the distribution of the mass ratio of the secondary ($\mu = M_2/M_B$) of the binary systems (see section 5.2.2). The parameter A is a free parameter representing the fraction in the IMF of binary systems with the right properties to give rise to SN Ia and is obtained by fitting the present time type Ia SN rate in the Galaxy.

The third integral in eq. (5.67) represents the mass restored by single stars with masses in the range $M_{Bm} - M_{BM}$; they can be either stars ending their lives as C-O white dwarfs or as type II SNe (those with $M > 8M_\odot$) The fourth integral refers to the material restored into the ISM by type II SNe while the first integral refers to the stars in the mass range $M_L - M_{Bm}$ with $M_L = m(t)$ being the minimum mass dying at the time t.

The quantity $Q_{mi}(t - \tau_m) = \sum_j Q_{ij}(m)X_j(t - \tau_m)$, where $Q_{ij}(m)$ is the production matrix as defined in eq. (2.71) and $X_j(t - \tau_m)$ is the abundance of the element j originally present in the star and later transformed into the element i and ejected, contains all the stellar nucleosynthesis. This formulation allows us to compute the evolution of the abundance of any chemical element, either those produced or destroyed (astration) inside stars such as D, ^{3}He, Li, Be and B.

The function describing the stellar lifetimes $\tau_m(m)$ is a necessary ingredient if one decides to integrate the complete equations (5.67). The time τ_{m2} is the lifetime of the secondary star in the binary system giving rise to a SN Ia, and represents the clock of the system. We give here some of the more widely adopted formulations for the stellar lifetimes.

Talbot and Arnett (1971) adopted a simple power-law in mass for the stellar lifetimes of the form:

$$\tau_m = 11.7 \cdot 10^9 m^{-2} \ yr \qquad (5.68)$$

for all masses.

Güsten and Mezger (1982) derived the following analytical formulas for the stellar lifetimes :

$$\tau_m = 5.0 m^{-2.70} + 0.012 \ (m \leq 8 M_\odot) \ Gyr \qquad (5.69)$$

$$\tau_m = 1.2 m^{-1.85} + 0.003 \ (m > 8 M_\odot) \ Gyr \qquad (5.70)$$

However, eq. (5.69) should be used only for stars with $m \geq 2 M_\odot$ since for lower mass stars it gives values for τ_m which are too small. The reason is that it was simply linearly extrapolated in the range of low mass stars.

Padovani and Matteucci (1993) adopted the following formula which gives more appropriate lifetimes for low mass stars :

$$\tau_m = 10^{\left[1.338 - \sqrt{1.790 - 0.2232 \cdot (7.764 - logm)}\right]/0.1116} - 9 \ Gyr \qquad (5.71)$$

for stars with $m \leq 6.6 M_\odot$, whereas for stars with $m > 6.6 M_\odot$ they used eq. (5.70).

Theis et al. (1992) adopted the following expressions, derived from Maeder and Meynet (1989) evolutionary tracks:

$$\tau_m = 1.2 \cdot 10^{10} m^{-2.78} \ yr, \ m < 10 M_\odot \qquad (5.72)$$

$$\tau_m = 1.1 \cdot 10^8 m^{-0.75} \ yr, \ m \geq 10 M_\odot \qquad (5.73)$$

Gibson (1997) adopted a more sophisticated approach by using the metallicity dependent lifetimes provided by new, extensive grids of stellar evolutionary tracks. He concluded that the metallicity effect on stellar lifetimes becomes evident only for low mass stars ($m \leq 3 M_\odot$), where the Z=0.001 lifetimes can be 30% $\rightarrow$ 40% smaller than the corresponding solar metallicity model. For a comparison between different stellar lifetime formulas see Figure 5.3.

5.2.1   THE MASS OF LIVING STARS

A very useful quantity in order to calculate the distribution of the dwarfs with metallicity is the mass (or the number) of living stars at any time t. In particular, the mass of living stars of mass m at the time t is defined as:

$$S_m(t) = \int_0^t \psi(t' - \tau_m) dt' \qquad (5.74)$$

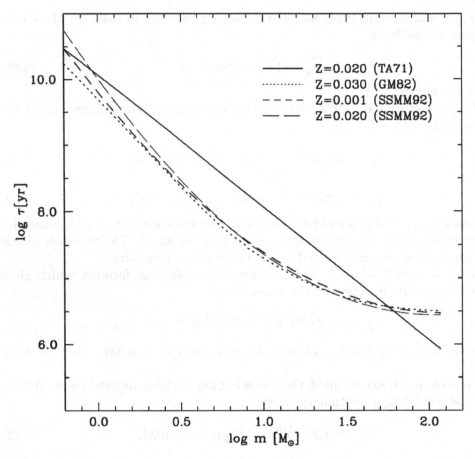

Figure 5.3. Different $\tau_m - m$ relations; TA71 (Talbot and Arnett 1971), GM82 (Güsten and Mezger 1982), SSMM92 (Schaller et al. 1992). From Gibson 1997, M.N.R.A.S. Vol. 290, 471; reproduced here by kind permission of Blackwell Science LTD (copy right 1997).

if $\tau_m \geq t$ and:

$$S_m(t) = \int_{t-\tau_m}^{t} \psi(t' - \tau_m)dt' \tag{5.75}$$

if instead $\tau_m < t$.

The total mass (number) of living stars at the time t is:

$$S(t) = \int_{m_L}^{m_U} S_m(t)\varphi(m)dm \tag{5.76}$$

where the IMF $\varphi(m)$ can be expressed either in number or in mass. With the quantity $S(t)$ one can build the cumulative distribution of stars in the solar neighbourhood (Figure 5.1). However, it is more meaningful to plot the differential distribution rather than the cumulative one since the former represents the variations of the star formation rate during the galactic lifetime whereas the latter gives us information only on the total amount of mass which has gone to form stars. Therefore the quantity to plot is $\frac{dS(t)}{dt}$ (see Figure 5.4). In Figure 5.4 we show the predictions of a numerical infall model concerning the differential distribution of the G-dwarfs in the solar neighbourhood.

It is worth noting that the observed metallicity has always been indicated by [Fe/H] because, since the original paper of Pagel and Patchett (1975) the stellar metallicity has been measured through the ultraviolet excess which is dominated by iron lines. On the other hand, the early chemical evolution models with instantaneous recycling approximation were comparing the predicted total metallicity Z with the observed [Fe/H]. This is not a correct procedure since Z is dominated by oxygen, which is an α-element and is produced on short timescales by massive stars, and therefore its abundance evolves quite differently from that of Fe which is produced both on short and long timescales (see chapter 1 and 6). The effect of comparing different abundances is shown in Figure 5.5 where we compare the predictions of a model without I.R.A. for the metallicity distribution of G-dwarfs either relative to Z or to Fe.

5.2.2   THE THEORETICAL SUPERNOVA RATES

We describe here some methods to calculate the supernova rates in the framework of chemical evolution models and we discuss such rates for our Galaxy.

Type II supernovae

The type II supernova rate is easy to define:

$$R_{SNII} = (1 - A) \int_{M_{up}}^{M_{BM}} \varphi(M)\psi(t - \tau_m)dM + \qquad (5.77)$$

$$+ \int_{M_{BM}}^{M_U} \varphi(M)\psi(t - \tau_m)dM$$

Where the extremes of the integral are M_{up}, namely the limiting mass for the formation of a degenerate carbon/oxygen core, which is defined in the range $6 - 8M_\odot$ according to the treatment of convection (see chapter 2), and M_U which is the largest considered mass. Normally, in chemical evolution models M_U is defined in the interval $40 - 100M_\odot$.

Type Ia supernovae

The type Ia supernova rate is less easy to compute and depends on the assumed model for the progenitors of type Ia SNe.

164

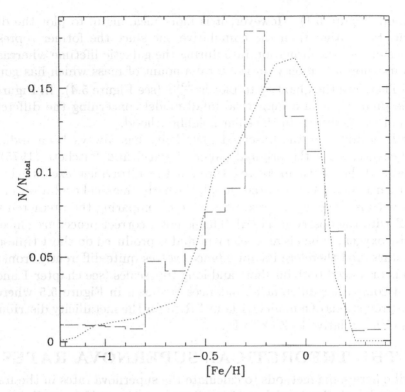

Figure 5.4. Differential distribution of G-dwarfs in the solar neighbourhood. *Dashed line*: the observational data from Rocha-Pinto and Maciel (1996). *Dots*: the predictions of the Chiappini et al. (1997) model which assumes infall of extragalactic material. From Chiappini et al. 1997, Ap.J. Vol. 477, 765; reproduced here by kind permission of C. Chiappini and the University of Chicago Press (copy right 1997).

In particular, if we assume the model of the C-O white dwarf plus the red giant, as in equation (5.67), the type Ia supernova rate can be written as:

$$R_{SNIa} = A \int_{M_{Bm}}^{M_{BM}} \varphi(m)[\int_{\mu_{min}}^{0.5} f(\mu)\psi(t - \tau_{m2})d\mu]dm \qquad (5.78)$$

where the function $f(\mu)$ is the distribution of the mass fraction of the secondary star in the binary system, namely $\mu = \frac{M_2}{(M_1+M_2)}$, with M_1 and M_2 being the primary and secondary mass of the system, respectively. This function is derived observationally and indicates that mass ratios close to the unity are favored. In the literature it has often been used the following expression:

$$f(\mu) = 2^{1+\gamma}(1 + \gamma)\mu^{\gamma} \qquad (5.79)$$

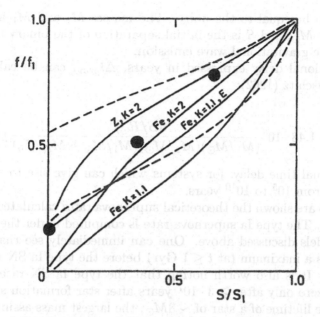

Figure 5.5. Cumulative distribution of G-dwarfs in the solar neighbourhood. *Dashed lines*: the observational data from Pagel and Patchett (1975). *Full points*: observational data from Bond (1970) and Schmidt (1963). The quantity $f = 10^{[Fe/H]}$ and S is the mass of stars with iron abundance $\leq f$. The values f_1 and S_1 refer to the values of these quantities at the Sun age. The label k on each curve refers to the assumed exponent of the star formation rate. The label Fe and Z refer to predictions relative to the real Fe abundance and to the generic metal abundance Z, respectively. Finally, the label E refers to a model with enriched infall. From this figure one can clearly see that predictions for the generic metal content Z are quite different from those for the Fe abundance, which is what is really observed. From Matteucci 1989, in "Evolutionary Phenomena in Galaxies", eds. J.E. Beckman & B.E.J. Pagel, p.297; reproduced here by kind permission of Cambridge University Press (copy right 1989).

for $0 < \mu \leq \frac{1}{2}$. The quantity γ has generally been used in the literature as a free parameter with a preferential value $\gamma = 2$. The lifetime τ_{m2} refers to the secondary star and this is because this star represents the clock of the system (see chapter 2).

If one adopts the double-degenerate model for the progenitors of supernovae Ia, the supernova rate can be computed in the following way (Tornambè and Matteucci 1987):

$$R_{SNIa} = Cq \int_{M_{min}}^{M_{max}} \varphi(M)[\int_{S_{min}}^{S_{max}} \psi(t - \tau_m - \Delta t_{grav})dlogS]dM \qquad (5.80)$$

where C is a normalization constant, $q = M_2/M_1 = 1$ is the ratio between the secondary and the primary mass which is generally assumed, for the sake of simplicity, to be equal to the unity. The masses M_2 and M_1 are defined in the interval 5-8 $M_\odot$, and S is the initial separation of the binary system at the beginning of the gravitational wave emission.

The gravitational delay expressed in years, Δt_{grav}, can be calculated from Landau and Lifschitz (1962):

$$\Delta t_{grav} = 1.48 \cdot 10^8 \frac{(S/R_\odot)^4}{(M_1/M_\odot)(M_2/M_\odot)(M_1/M_\odot + M_2/M_\odot)} \, yr \qquad (5.81)$$

This gravitational time delay, for systems which can give rise to supernovae of type Ia varies from 10^5 to 10^{10} years.

In Figure 5.6 are shown the theoretical supernova rates calculated for the solar neighbourhood. The type Ia supernova rate is computed under the two different progenitor models discussed above. One can immediately see that the type II SN rate reaches a maximum (at $t \leq 1$ Gyr) before the type Ia SN rate (between 2 and 4 Gyr). It is also worth noting that the type Ia SN rate starts to be different from zero only after $\geq 3 \cdot 10^7$ years after star formation started, which corresponds the lifetime of a star of $\sim 8M_\odot$, the largest mass assumed for SN Ia progenitors in the single degenerate model. This is the minimum timescale for the appearance of type Ia SNe after the beginning of star formation since the double degenerate model as well as more recent models for progenitors of type Ia SNe, with lower mass secondary stars, imply longer timescales (see chapter 2).

The theoretical supernova rates are quite important in computing the chemical evolution of galaxies since their functional form and absolute values are fundamental in determining abundance ratios between elements formed in different supernova types and absolute abundances. For example the delay in the occurrence of type Ia SNe relative to SN II is very important to explain the behaviour of α-elements and Fe, as we will see in the following chapters.

5.2.3 TIMESCALES FOR TYPE IA SN ENRICHMENT

In the past years there has been a lot of discussion on the role of supernovae of different type in the chemical enrichment of galaxies. In particular, on the fact that the observed trend of the $[\alpha/Fe]$ vs. $[Fe/H]$ (see chapter 1) can be explained by the different roles of SN II and Ia in the galactic chemical enrichment. In this framework, the plateau observed at low metallicities is due to the almost constant α/Fe ratio produced by massive stars which explode as type II SNe on short timescales, while the subsequent decline is due to the contribution of SN Ia which explode on longer lifetimes. This interpretation of the observed abundance ratios was first suggested by Tinsley (1980) and then by Greggio and Renzini (1983b) but the first detailed calculation without I.R.A. of the contribution of type Ia SNe to the galactic enrichment is by Matteucci and Greggio (1986). They

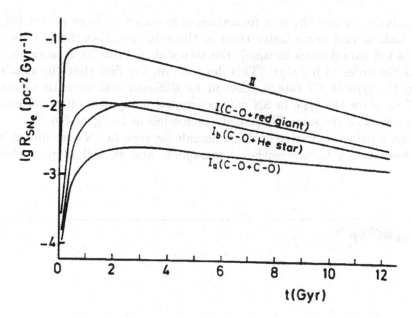

Figure 5.6. Predicted supernova rates as functions of time in the solar neighbourhood. The units of the SN rates are number of SNe $pc^{-2}Gyr^{-1}$ and the time is in Gyr. In the figure are indicated the predictions by different models of type Ia SN progenitors (C-O plus C-O and C-O plus RG). The type Ib SN rate is also indicated: it has been calculated by assuming that the progenitors of such SNe are binary systems made of a C-O white dwarf plus an He-star, as discussed in Tornambè and Matteucci (1987). From Matteucci 1991, in "Frontiers of Stellar Evolution" A.S.P. Conf. Series Vol. 20, 539; reproduced here by kind permission of the Astronomical Society of the Pacific (copy right 1991).

showed that it was possible to reproduce the observations of [O/Fe] vs. [Fe/H] in galactic stars in the framework of a model such as that described in 5.2 with a constant IMF which was assumed to be that of Scalo (1986). They pointed out that, in a good model for the chemical evolution of the solar region, the turning point for the [O/Fe], coinciding with the end of the halo phase (namely [Fe/H]~ -1.0 dex), occurs at a galactic age of 1-1.5 Gyr. This timescale can, in principle, be taken as the timescale for the formation of the halo because of this coincidence, which is due to the fact that at that epoch the iron produced by type Ia SNe starts to become important, thus lowering the [O/Fe] ratio, which decreases since then down to the solar value ($[O/Fe]_\odot$=0). This timescale can therefore be considered as the typical timescale for the iron enrichment and many authors have adopted this conclusion; in particular, it is generally assumed that the typical timescale for type Ia SN enrichment is 1 Gyr. However, it should be noted that the timescale for type Ia SN enrichment is not universal, in the sense that it depends not only on the lifetimes of the assumed progenitor model for type Ia SNe but also on the assumed history of star formation. For example, in elliptical galaxies, where the star formation is assumed to have proceeded in a burst-like fashion and much faster than in the solar neighbourhood (where we know it has not varied much in time), the timescale for the SN Ia enrichment is shorter, of the order of 0.3 Gyr. This depends on the fact that the maximum reached by the type Ia SN rate is different for different star formation rates. In Figure 5.7 we show the type Ia SN rate computed for typical elliptical galaxies in the framework of the same SN progenitor models as in figure 5.6.

Therefore, a better definition of the timescale for type Ia SN enrichment is the *time at which the SN Ia rate reaches a maximum*, and we will adopt it through the book.

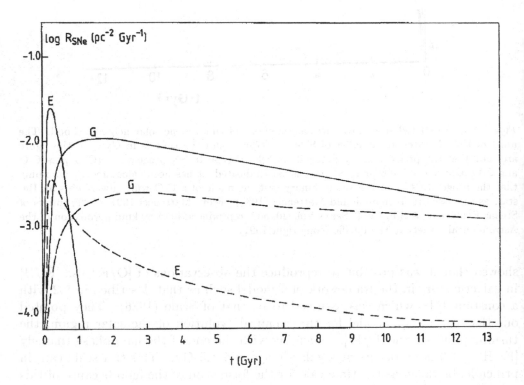

Figure 5.7. Predicted supernova (Ia and Ib) rates as functions of time for elliptical galaxies (labeled E) and the solar neighbourhood (labeled G). The units of the SN rates are number of SNe $pc^{-2}Gyr^{-1}$ and the time is in Gyr. The dashed curves represent the type Ia SN rates while the continuous ones represent the type Ib SN rates. The type Ib supernova rate has been calculated by assuming that the progenitors of such supernovae are binary systems made of a C-O white dwarf plus an He-star, as discussed in Tornambè and Matteucci (1987). From Tornambè and Matteucci 1987, Ap.J. Vol 318, L25; reproduced here by kind permission of A. Tornambè and the University of Chicago Press (copy right 1987).

5.2.4 THEORETICAL NOVA RATES

Theoretical nova rates have been computed in models of chemical evolution in order to account for their production of ^7Li and ^{13}C and ^{15}N (e.g. D'Antona and Matteucci, 1991). These authors assumed that the rate of formation of nova systems is proportional to the rate of formation of white dwarfs, namely:

$$R_{nova}(t) = \alpha \int_{0.8}^{8} \psi(t - \tau_m - \Delta t)\varphi(m)dm \qquad (5.82)$$

where α is a free parameter representing the fraction of white dwarfs which belong to binary systems of the type giving rise to nova systems, and it is fixed by reproducing the observed present time nova rate (outburst rate) after assuming that each nova system would suffer a number $n \simeq 10^4$ outbursts during its lifetime. The quantity $\Delta t \sim 1$ Gyr represents the time delay between the formation of the WD and the first nova outburst. For the sake of simplicity, it is assumed that all the nova outbursts occur at the same time. In Figure 5.8 we show the predicted birthrates of white dwarfs and nova systems as functions of the galactic lifetime, computed according to eq. 5.82.

The nova rate as well as the supernova rates depend not only on their assumed progenitors (namely, stellar mass ranges, binary or single stars, IMF, common-envelope efficiency) but also on the assumed star formation history. In Figure 5.9 the evolution of nova rates under the assumption of different SFRs is shown.

5.3 CHEMO-DYNAMICAL MODELS

The term *chemo-dynamical* was introduced first by Burkert and Hensler (1988; 1989) to describe models which combine hydrodynamical and chemical evolution calculations. However, chemo-dynamical models had already been extensively studied in the pioneering work of Larson (1969, 1974, 1975, 1976), who showed first that gas dynamics and energy dissipation play a fundamental role in galactic evolution.

One of the models of this kind is described in Theis et al. (1992). This model takes into account different gas phases (hot and cold) by means of heating and cooling processes. These two phases represent the hot tenuous intercloud medium with temperatures above 10^4K and the cloudy gas with temperatures below 10^4K, respectively. The heating processes are represented by the energy injected from the stars into the ISM through stellar winds and supernova explosions, while the cooling law follows the prescriptions of Hensler (1988). The dynamical evolution of both stars (with Boltzmann moment equations) and gas (with Eulerian equations), which are dynamically coupled through self-gravity, are followed and the gas metallicity (Z) is also taken into account. The I.R.A. is relaxed so that stellar lifetimes are considered, although consideration of stellar lifetimes is important only if one considers the evolution of single elements and,

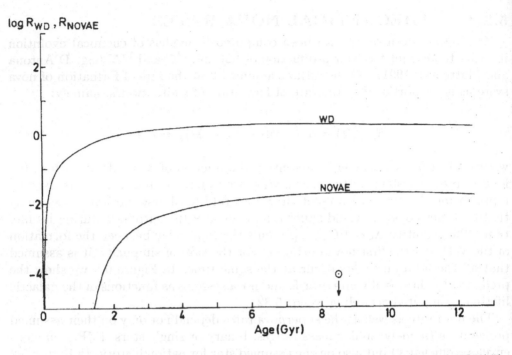

Figure 5.8. Predicted birthrates of WDs (labeled WD) and nova systems (labeled NOVAE) as functions of the galactic age. The parameter $\alpha = 0.01$ is chosen to fit the present time observed rate of nova outburst. The birthrates are expressed in units of number $pc^{-2}Gyr^{-1}$. From D'Antona and Matteucci 1991, A&A Vol. 248, 62; reproduced here by kind permission of Springer Verlag (copy right 1991).

in particular, of those elements produced on timescales comparable to the Hubble time. On the other hand, I.R.A. is a reasonable approximation to compute the evolution of the global metal content, since Z is dominated by oxygen which is produced on timescales much shorter than the galactic lifetime (see chapter 6). The star formation rate follows essentially the Schmidt (1959) law but with the inclusion of a dependence on the gas temperature. A dark matter halo is not considered under the justification that the model refers to elliptical galaxies where there is no evidence of dark matter on galactic scales (e.g. Binney et al. 1990; van der Marel et al., 1990) However, it is likely that dark matter in ellipticals is contained in heavy but diffuse halos (Carollo et al. 1995), and the inclusion of such halos in chemical evolution models does not have a negligible effect (Matteucci and Gibson, 1995). The initial conditions of Theis et al.'s model consists in a purely gaseous, isotropic and homogeneous system in virial equilibrium and the calculations are performed either in 1 or 2 dimensions. This chemo-dynamical approach applied to ellipticals has basically confirmed Larson's results on the necessity of considering dissipative processes in galaxy evolution

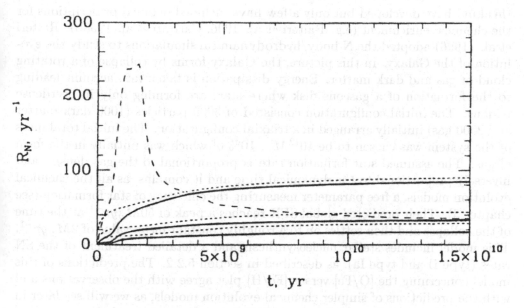

Figure 5.9. Evolution of the rate of novae (expressed in units of yr^{-1}) in the constant star formation rate model. Pairs of curves correspond to (top to bottom) different exponents of the initial distribution of binaries (-1,0,1 respectively) as a function of their mass ratio. In each pair, the short dashed line describes the case referring to a common-envelope parameter α_{CE} =0.5, whereas the thick line refers to $\alpha_{CE} = 1$ (the common envelope parameter is a measure of the efficiency of the common envelope process, see Iben and Tutukov 1984). The long-dashed line describes the evolution of the nova rate for the case in which a 10^9 yr initial burst of star formation is followed by a stage of constant star formation for another $14 \cdot 10^9$ yr. From Yungelson et al. 1997, Ap.J. Vol. 481, 127; reproduced here by kind permission of M. Livio and the University of Chicago Press (copy right 1997).

and indicated that a multi-phase approach for the treatment of the gas is required. As a consequence of this multi-phase approach, interesting results have been obtained for dwarf galaxies, where the chemo-dynamical model predicts a self-regulating mechanism for star formation with alternative burst and wind phases (Hensler and Burkert, 1990).

More recently Steinmetz and Müller (1994; 1995) investigated the viability of applying the smooth particle hydrodynamics (SPH) technique to the chemical evolution of galaxies. In this kind of approach, gas, stars and dark matter are treated as particles and their evolution is followed by means of N-body algorithms. The spatial and temporal evolution of each particle is followed separately from all the others; this leads to the advantage of studying the chemical evolution on a relatively small scale with the consequent creation of inhomogeneities. In this first study, the authors computed the chemical enrichment due only to SN II since this allowed them to adopt I.R.A. In the following years other studies of this kind have developed but only a few have included detailed prescriptions for the chemical enrichment (e.g. Raiteri et al. 1996; Carraro et al. 1998b). Raiteri et al. (1996) adopted the N body/hydrodynamical simulations to study the evolution of the Galaxy. In this picture, the Galaxy forms by collapse of a rotating cloud of gas and dark matter. Energy dissipation is taken into account leading to the formation of a gaseous disk where stars are forming only in overdense regions. The initial configuration consisted of 3000 particles (1000 dark matter and 2000 gas) initially arranged in a triaxial configuration. The initial total mass of the system was chosen to be $10^{12} M_\odot$, 10% of which was initially in the form of gas. The assumed star formation rate is proportional to the gas density and inversely proportional to the dynamical time and it contains, as all the chemical evolution models, a free parameter measuring the efficiency of star formation (see chapter 3). Their predicted global SFR reaches a peak of $50 M_\odot yr^{-1}$ at the time of the collapse and then settles on to a roughly constant mean value of $2 M_\odot yr^{-1}$. This model includes stellar nucleosynthesis and a detailed treatment of the SN rates (type II and type Ia), as described in section 5.2.2. The predictions of this model concerning the [O/Fe] versus [Fe/H] plot agree with the observations and with the predictions of simpler chemical evolution models, as we will see later in chapter 6, although the predicted spread is larger than observed.

The chemo-dynamical approach to galactic evolution certainly seems to be the most promising one, although at the present time the number of free parameters contained in this kind of models is even larger than in purely chemical models, and a very fine tuning of the various parameters is necessary to reproduce real galaxies. Moreover, the results produced by these models are more difficult to interpret than simple chemical results and so far they have only confirmed some of the most obvious observational constraints, such as the age -metallicity relation but the modelling is continuously improving and we expect very important results in the near future.

Chapter 6

FORMATION AND EVOLUTION OF THE MILKY WAY

The problem of understanding the formation of the Galaxy is part of the problem of explaining galaxy formation in general. In particular, we should try to understand the relative importance of mergers and dissipative collapse in the formation of all galaxies, and whether the observational information about our Galaxy is enough to explain the timescales and the mechanisms of its formation and if this can be in agreement with currently popular hierarchical merger models. As we will see in what follows, it is from the study of chemical and dynamical properties of the different stellar populations that we can gain information on galactic formation and evolution. In fact, the chemical composition in the atmospheres of main-sequence stars represents, in most cases, the chemical composition of the interstellar medium (ISM) from which the stars were formed. Besides, kinematical properties of the stars contain information about the conditions of the gas at the time of star formation, since the dynamical relaxation time is sufficiently long compared to the age of the Galaxy. Thus the metal abundances and the kinematics of the stars represent important information concerning the history of the chemical and dynamical evolution of the Galaxy. An extensive review of galactic structure surveys can be found in Majewski (1993).

6.1 ABUNDANCES AND KINEMATICS

The landmark paper of Eggen, Lynden-Bell and Sandage (1962) (hereafter ELS) was the first attempt to understand the formation and evolution of our Galaxy. From a study of a kinematically selected sample of high velocity stars, they found a remarkable correlation between chemical abundance (indicated by [Fe/H]) and orbital eccentricity e, in the sense that stars with the lowest metallicity are invariably moving in highly elliptical orbits. As the average $< [Fe/H] >$ would be expected to increase with time, as a consequence of the progressive chemical enrichment of the gas, stars with the lowest [Fe/H] are, on average, the oldest. They also found a correlation between abundance and motion of stars perpendicular to the galactic plane, evidenced by a continuous decrease of the perpendicular velocity with increasing [Fe/H]. To explain these relations they proposed that the Galaxy collapsed from a protocloud to a thin-disk on a timescale of a few times 10^8 years, with progressive chemical enrichment taking place as the collapse proceeded. On the other hand, if the concentration of mass occurred slowly either in a quasi-static contraction or merging/accretion of separate fragments, some fraction of metal poor stars would be expected to exhibit nearly circular orbits.

This model was subsequently criticized mainly because of the selection effects present in their data, i.e. given the data available to them one would not expect the sample to contain low abundance, low orbital eccentricity objects even if they existed, since they would have been absent from the adopted high velocity catalog. In addition, the proposed simple model did not account for the fact that almost half of the halo stars have *retrograde orbits*. This fact led Larson (1969) to consider models of clumpy and turbulent protogalaxies with collapse times that sometimes exceeded 1 Gyr.

Searle and Zinn (1978) (hereafter SZ) proposed a different view of the formation of the Galaxy as a result of their studies of the globular cluster system. Firstly, they found no gradient in metallicity in the globular clusters at galactocentric distances larger than 10 kpc. Secondly, the HB morphology in globular clusters appears to be a function of galactocentric distance. This was attributed to age differences, implying that globular clusters in the outer halo have an age spread of a couple of billion years, whereas those of the inner halo have a spread in age of less than 1 Gyr. They also claimed that for halo abundances ($[Fe/H] < -1.0$ dex) there is no dependence of the kinematics on abundance for subdwarf samples when all the velocity components are considered. Therefore, they proposed a major modification of the ELS picture in which, in addition to a central collapse, they envisaged gaseous fragments in the outer halo which remained distinct and experienced their own star formation and chemical evolution over a relatively long period before reaching equilibrium with the inner galactic regions. In this way, one should expect no gradient and no correlation between kinematics and abundance, as well as an age spread for the globular cluster population of the outer halo.

Yoshii and Saio (1979), analyzing a large sample of stars, concluded that the kinematics can be better explained by a slow rather than a fast collapse model, with the collapse lasting $\sim 3 \cdot 10^9$ years. They did not find a precise correlation between [Fe/H] and e below [Fe/H] < -1.0 dex.

Later, Yoshii (1982) identified a third major stellar component of the Galaxy which was later confirmed and named *thick disk* by Gilmore and Reid (1983) and Gilmore and Wyse (1985). This component, which possesses some properties of the disk subsystem of globular clusters discussed by Zinn (1985), was excluded by ELS since they used catalogs containing only samples representative of the kinematic extremes.

In the following years Sandage (1987) and Sandage and Fouts (1987) confirmed the existence of a large population of intermediate velocity stars, with a metallicity distribution intermediate between that of the old thin disk (e-folding scale height of $\simeq$ 270 pc) and the extended halo (e-folding scale height $\simeq$ 3.2 kpc) having an e-folding scale-height of 920 pc. The mean metallicity found for the thick-disk stars is $< [Fe/H] >_{thick}= -0.6$ dex, whereas that of the halo is $< [Fe/H] >_{halo}= -1.5$ dex and that of the thin disk $< [Fe/H] >_{thin}= -0.1$ dex. They identified this component as the thick-disk of Gilmore, Reid and Wyse. The main conclusion of Sandage was that the addition of the thick-disk to the thin-disk and halo changes only the details of ELS picture but not its essential substance. The conclusions of Sandage were again in contrast with those of SZ and Norris (1986), who claimed no correlation between metallicity and kinematics below [Fe/H]< -1.0 dex, whereas Sandage found a continuous correlation. If Sandage is correct, the chemical evolution of our Galaxy can be thought of as a continuum process with the collapse to enrichment rate ratio varying as a function of time. In the beginning the collapse must have been rapid with a very small energy dissipation rate, after which the dissipation increased thus decreasing the collapse rate. At the same time, the metal enrichment increased and thus the ratio of the collapse to enrichment rate levelled off producing higher metallicity per unit mass in stars in the flattened structures (thick and thin-disk). In this way, we can explain how each galactic component has a different average metallicity, chemical gradient and metallicity distribution. This scenario is illustrated in Figure 6.1.

On the other hand, in agreement with the previous study of Norris, Ryan and Norris (1991) concluded, from a kinematical study of metal poor stars, that their data are not consistent with a model in which the halo formed from star formation in a dissipating collapsing cloud, unless the kinematics have since been modified by unstable orbits or violent relaxation. They suggested a picture in which the halo stars formed in numerous independently evolving clouds, similar to Larson's (1969) earlier suggestion.

A different scenario, *the hot Galaxy picture*, has been proposed by Berman and Suchkov (1991) (hereafter BS). As an alternative to the ELS and SZ pictures, BS suggested that the large energy release triggered by the initial burst of galactic star creation quickly inhibited further star formation. A strong wind driven by the initial intense burst of star formation ejected enriched material from the proto-Galaxy into intergalactic space. Subsequently, the remainder of the proto-Galaxy, after a delay of a few gigayears contracted and cooled to form the major stellar components observed today.

Unfortunately, at present we are not yet able to distinguish which of these is the correct scenario, mainly because, as Gilmore et al. (1989) have pointed out, the presence or absence of a correlation between rotation velocity and stellar metallicity does not really allow one to distinguish between fast and slow collapse models of the Galaxy. Therefore, before trying to reach any conclusion about the Galaxy formation we should discuss another independent clock which is given by purely chemical arguments, as will be seen in the following discussion.

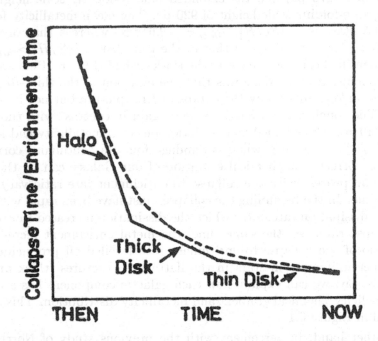

Figure 6.1. The formation of the Milky Way. Schematic representation of the variation with time of the ratio of the collapse rate to the metal enrichment rate. The continuous process leading to a smooth merging of the kinematic, spatial and metallicity components of the Galaxy can be approximated by straight line segments as if there are three discrete population components. From Sandage 1987, in "The Galaxy", ed. G. Gilmore and B. Carswell, NATO ASI Series, p.321; reproduced here by kind permission of Reidel, Dordrecht (copy right 1987).

6.2   THE GLOBULAR CLUSTERS AGES AND THE FORMATION OF THE HALO

Age differences among globular clusters can represent a very important constraint to the mechanism of the formation of the galactic halo. In order to make this practical, one needs to know the relative ages of globular clusters rather than their absolute ages.

On the other hand, it is very important to know the absolute ages of globular clusters since they are likely to be the oldest objects in the universe. Unfortunately, it is a very difficult task to measure the absolute ages of globular clusters: one needs to determine the mass of those stars at the MS Turn-Off (MSTO), given the observed brightness and colour of the MSTO. Then the observed magnitude and colour of the MSTO are finally converted to absolute magnitude M_V and color $(B - V)_o$, respectively, using knowledge of the distance to and the reddening towards the cluster. These quantities finally are converted into M_{bol} and T_{eff} through the use of stellar models adopting the appropriate abundances ([Fe/H], [α/Fe] and the helium mass fraction Y). This is a very complex procedure with many free parameters. Among other things, one should note that the metallicity of globular clusters, always indicated for historical reasons with [Fe/H], in many cases has nothing to do with iron but reflects the abundances of other elements such as calcium. This is a very important point to remember since the α-elements (O, Mg, Ne, Si, Ca, S, Ti) behave quite differently from Fe, showing in fact a marked overabundance in metal poor objects, including globular clusters (see chapter 1). Therefore, one should not confuse the various elements.

Another point of possible confusion is the adoption of overabundances of certain elements, rather than others, relative to Fe. For example, many stellar evolution studies often referred to an overabundance in the CNO elements relative to Fe ($[CNO/Fe] > 0$); this is not correct since C and N are not found to be overabundant relative to iron in the same way as oxygen, as we discussed already in chapter 1.

The question then is, *how do uncertainties in the input parameters propagate through the determination of cluster ages?*

Bolte and Hogan (1995) quantified these uncertainties and concluded that every input parameter, besides stellar colours which are the most uncertain predicted quantities in stellar models, contribute significantly to the uncertainties in the derived ages when using the MSTO colour as an age indicator. On the other hand, if the MSTO luminosity is used to derive cluster ages, the only significant source of error is in cluster distance determinations.

Gratton et al. (1997b) used new distance determinations based on Hipparcos subdwarf parallaxes and derived ages for globular clusters. Their best value for the age of the oldest clusters is:

$$Age = 11.8^{+2.1}_{-2.5} \ Gyr \tag{6.1}$$

where the error bar is the 95% confidence range. They argued that if one allows for ~ 0.5 Gyr for cluster formation, this age estimate is then compatible with an Einstein-de Sitter Universe with $H_o \leq 64$ km sec^{-1} Mpc^{-1} or with a flat Universe with $\Omega_m = 0.2$ if $H_o \leq 83$ km sec^{-1} Mpc^{-1}. More recently, Pont et al. (1998), also using Hipparcos parallaxes for subdwarfs, deduced an age of 14 Gyr for the globular cluster M92.

From the theoretical point of view, the distance scale to globular clusters derived recently through new horizontal branch models (Mazzitelli et al. 1995; Caloi et al. 1997) also suggested an age for the globular clusters of $\simeq 12$ Gyr (see also chapter 7).

However, for the purpose of our discussion we are more interested to relative ages. There are two principal techniques for deriving relative ages:

- the luminosity difference (ΔV) between the horizontal branch and the MSTO (Iben and Faulkner, 1968), and

- the color difference, typically $\Delta(B-V)$ or $\Delta(V-I)$ between the turn-off and the base of the red giant branch (e.g. Vandenberg et al., 1990).

Both these techniques are based on the fact that the differences in luminosities and colours can be expressed as a function of the cluster age and both have advantages and weaknesses. In particular, the first method (based on luminosity) is operationally more difficult to apply than the second method (based on colour) and suffers from uncertainties about the dependence of the HB absolute magnitude on metallicity. The colour method, on the other hand, suffers mainly from the uncertainties related to the theoretical RGB temperature, which in turn is very sensitive to the physics adopted in computing the stellar evolutionary tracks. Relative ages derived with this method, which is the one most largely adopted at the present time, seem to indicate that most of the galactic globular clusters are old and coeval (in particular those with [Fe/H] < -1.2 dex), whereas intermediate and metal-rich clusters ([Fe/H] $\geq$ -1.2 dex) appear to be younger by $\sim$ 15% of the age of the Universe. Moreover they indicate that there is no evident age-distance relation at least up to 20 kpc from the galactic centre (Rosenberg et al. 1999). The uniformity of the ages of the inner clusters (they do not differ by more than 0.5 Gyr) suggests that the initial burst of star cluster formation was a global, rapid phenomenon that occurred over a huge volume. Therefore, most of the halo population of globular clusters should be coeval with a fraction of the more metal rich clusters showing ages which are perhaps 2-3 Gyr younger.

It is worth noting the existence of a third method, developed by Lee et al. (1994), which uses synthetic HB models to analyze the distribution of stars on the HB of globular clusters. With this technique it was found that the different morphology of the HB stars in clusters at different galactocentric distances is best explained by age differences. In particular, for clusters at galactocentric distances $R < 8$ kpc the ages seem to be 2 Gyr older than for clusters with $8 < R < 40$ kpc which, in turn, seem to be 2 Gyr older than clusters at $R > 40$ kpc. Moreover, the inner halo appears to be more uniform in age than the outer halo.

Buonanno et al. (1994) showed that four globular clusters are unequivocally younger by 3-5 Gyr than others of similar metallicity and they are all at galactocentric distances larger than 20 kpc, and concluded that it is possible that they have all been accreted by the Milky Way since they are on similar orbits.

Although all of these age determinations are still uncertain, the results discussed before seem to suggest that the formation of galactic globular clusters started at the same time over the halo, at least inside 20 kpc from the galactic center, and lasted for 1-2 Gyr. On the other hand, the outermost halo (R> 20 kpc) probably formed on a longer timescale than the inner halo by dissipational collapse (*inside-out* formation) and/or some external globular clusters might have been accreted.

6.3 ABUNDANCE RATIOS AND GALAXY FORMATION

An independent way of deducing the timescales for the formation of the various galactic components is to look at the evolution of abundance ratios determined for stars in the solar vicinity. In recent years, a great deal of observational work has been devoted to the study of the chemical abundances in solar neighbourhood stars, as we have reviewed in chapter 1.

A particularly useful abundance ratio is that between α-elements and Fe since the α-elements (i.e. those synthesized from α particles such as O, Ne, Mg, Si and S) are almost entirely produced by massive short living stars ($M > 10 M_\odot$ type II SNe) with lifetimes in the range 10^6-10^7 years, whereas Fe is produced partly by type II SNe and mostly by type Ia SNe, which are believed to originate from white dwarfs in binary systems, with lifetimes ranging from several 10^7 to 10^{10} years and more (see chapter 2). Because of the time delay in the bulk of iron production relative to α-elements, the abundance ratio [α/Fe] can be used as a cosmic clock. Let us take oxygen as representative of α-elements; one can infer the timescale for the formation of the halo by looking at the observed [O/Fe] vs. [Fe/H] relation for solar neighbourhood stars. This relation shows a sort of plateau extending roughly up to [Fe/H]= −1.0 dex (which is also the canonical transition metallicity from halo to disk stars) and then the [O/Fe] ratio declines until it reaches the solar value (see Figure 6.2). If we have a galactic chemical evolution model following the evolution of single chemical abundances, such as that described in chapter 5, and we are able to reproduce the observed relationship, then we know at which time the gas has reached the metallicity [Fe/H]= −1.0 dex and this is the timescale for the formation of the halo, under the assumption that the various galactic components formed in a continuous process. The theoretical [O/Fe] vs. [Fe/H] relation depends on the assumed SN progenitors, stellar nucleosynthesis and star formation history. For this reason, the observed abundance patterns, when compared to theoretical models, can constrain SN progenitors, nucleosynthesis and star formation at the same time. Successful chemical evolution models suggest this timescale to be not longer than 1-1.5 Gyr (see Figure 6.2). In particular, Matteucci and François (1992) tried to reproduce the change in slope of [O/Fe] vs. [Fe/H] relation at [Fe/H] $\simeq$ -1.7 dex, suggested by Bessel et al. (1991). As we have seen in chapter 1, this change in slope or *knee* in more recent data occurs at [Fe/H] $\sim$ -1.0 dex, but this difference does not change significantly the conclusion for the halo timescale. In this framework, the reason for the existence of such a knee is the fact that it coincides with the timescale of type Ia SN enrichment as defined in chapter 5, namely the maximum in the type Ia SN rate. What is still puzzling is the coincidence between the end of the halo phase and the timescale of type Ia SN enrichment.

It should be mentioned that Edmunds et al. (1991) proposed an alternative way of interpreting the [O/Fe] vs. [Fe/H] behaviour by assuming that only SN II would produce iron and that the Fe yield would depend on metallicity, showing a secondary-like behaviour. This suggestion, although formally correct, does not seem very realistic since the envelopes of SN Ia as well as their light curves establish the presence of substantial amounts of Fe. However, the authors themselves proposed some tests to see if this *pseudo-secondary* model has to be preferred to the *time-delay* model. For example, one suggested test was to look at the bulge stars. In the time-delay model one expects that, owing to the fast evolution of the bulge, high [O/Fe] ratios would persist at high [Fe/H], as shown by Matteucci and Brocato (1990) and in Figure 6.3, whereas the pseudo-secondary model does not predict any difference relative to the prediction for the solar vicinity. In other words, the pseudo-secondary model would predict universal abundance ratios as functions of metallicity irrespective of the different star formation histories. As discussed in chapter 1, the data on bulge stars seem to favor the time-delay model. A simplified version adopting I.R.A. of the time-delay model was presented by Pagel (1989) who assumed a constant delay to account for the Fe production by type Ia SNe (see chapter 5). His predictions are shown in Figure 6.4. However, in this model the disk is assumed to have formed out of the gas lost from the halo, an hypothesis which does not seem likely, as we will discuss in the following.

Matteucci and François (1992) presented several models assuming different star formation laws and concluded that the data are best reproduced by a model where the star formation rate depends almost linearly upon the surface gas density. On the other hand, if star formation has a quadratic dependence on the surface gas density (Schmidt law) then this time scale is only $3 \cdot 10^8$ years and the change in slope in the predicted [O/Fe] vs. [Fe/H] relation occurs at [Fe/H] > -1.0 dex, at variance with the observations. Therefore, it is extremely important to have detailed abundance ratios involving oxygen and other α-elements especially in halo stars to impose constraints on the star formation history. The timescale of 1-1.5 Gyr is longer than that inferred by ELS but is in agreement with the estimated duration of the halo phase from globular cluster ages.

A large age spread was suggested also for halo field stars (subdwarfs) (Schuster and Nissen, 1989). These authors found an age-metallicity relation for halo stars indicating a difference of $\simeq$ 3 Gyr passing from [Fe/H]=-3.0 to -2.0 dex. If this result is real and is not affected by systematic errors it favours a coherent very slow dissipative collapse of the halo. However, a timescale of the order of 3-4 Gyr is at variance with that deduced from the [O/Fe] vs. [Fe/H] ratio since in order

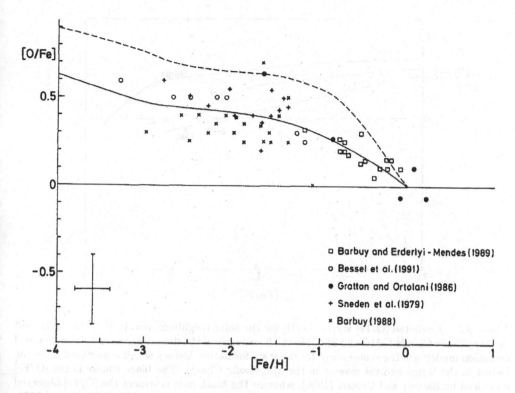

Figure 6.2. Observed [O/Fe] versus [Fe/H] for the solar neighbourhood (data sources are in the figure) compared with the predictions of Matteucci and François (1992) (solid curve). This model implies that [Fe/H]=-1.0 dex is attained after 1.5 Gyr from the beginning of star formation. From Matteucci and François 1992, A & A Vol. 262, L1; reproduced here by kind permission of Springer Verlag (copy right 1992).

to have such a slow halo collapse there should be a much more restricted region of constant [O/Fe] for halo stars. In fact, if the halo collapse time is longer than 3 Gyr then SN Ia had time to restore the bulk of Fe, since the characteristic timescale for this restoration is $\simeq$ 1 Gyr for the solar neighbourhood, and the change in the slope in the observed [O/Fe] should occur at much lower metallicities than observed([Fe/H]=-1.0 – -1.5 dex), as is shown in Figure 6.3. However, Marquez and Schuster (1994) found that there is not only one age-metallicity relation in halo stars as found before, but the inner part of the halo is older and with less age spread than the outer one, suggesting again that the halo formed from *inside out*.

Wyse and Gilmore (1993) suggested that an aspect of the [O/Fe] vs. [Fe/H] relation which might be in contrast with the *chaotic scenario* of SZ, is the small spread around [Fe/H]=-1.0. In fact, if the halo or part of it were the result of random evolution of independent clouds, we should perhaps expect a larger spread

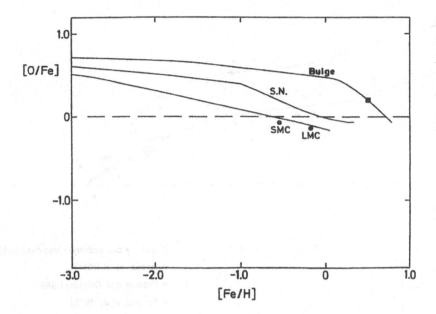

Figure 6.3. Predicted [O/Fe] versus [Fe/H] for the solar neighbourhood (S.N.), the Bulge and the Magellanic Clouds (LMC and SMC) in the framework of the time-delay model. The chemical evolution models are the same except for the stars formation history which is assumed to be the fastest in the Bulge and the slowest in the Magellanic Clouds. The black square is the [O/Fe] measured by Barbuy and Grenon (1990), whereas the black dots represent the [O/Fe] observed in the two Clouds by Russel et al. (1988) and Spite and Spite (1990). From Matteucci 1991, in "SN1987A and Other Supernovae", eds. I.J. Danziger and K. Kjär, p. 703; reproduced here by kind permission of European Southern Observatory (copy right 1991).

in the [O/Fe] ratio in halo stars, especially because of the different metallicities at which the change in the [O/Fe] ratio would occur in the different clouds. The change in the [O/Fe] is indeed very sensitive to the star formation history, as is clearly illustrated in Figure 6.3.

These discrepancies could be eliminated if we think not in terms of a unique timescale for the halo collapse but assume that the inner halo evolved more rapidly than the outer halo, and that the chemically derived timescale refers mostly to the inner halo, due to the fact that the halo stars observed in the solar region are mostly those born inside the solar circle (Grenon 1987). This agrees also with the fact that the observed spread in the age of globular clusters refers only to objects belonging to the outer halo, and with what was suggested by Zinn (1993), namely that the inner old halo clusters could have formed during a collapse not unlike the one proposed by ELS, whereas the younger outer halo clusters could have formed in satellite systems that were accreted by the Galaxy. In the best model of Matteucci and François (1992) collapse timescales increasing with galactocentric distance are assumed. As a consequence, they predicted that

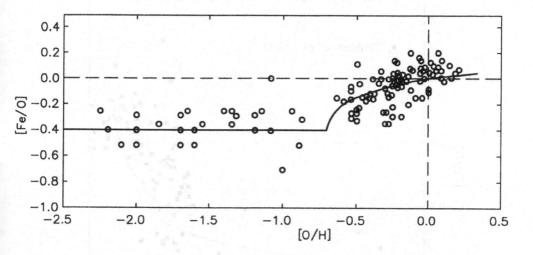

Figure 6.4. Predicted [Fe/O] versus [O/H] for the solar neighbourhood in the framework of the time-delay model of Pagel (1989). Note the completely flat [Fe/O] predicted for the halo as a consequence of the I.R.A. Adapted from Pagel 1994, in "Galaxy Formation and Evolution", eds. C. Munoz-Tunon & F. Sanchez; reproduced here by kind permission of B.E.J. Pagel and Cambridge University Press (copy right 1994).

both halo and disk take longer to form in the outermost regions. In particular, they estimated, from the chemical argument, that the extreme outer halo should have formed in 3-4 Gyr as compared to the 1-1.5 Gyr predicted for the inner halo.

Another observational finding which could favor the *inside-out* scenario is the fact that Nissen and Schuster (1997) identified some metal rich halo stars (-1.3 $\leq [Fe/H] \leq -0.5$ dex) with both high and low [α/Fe] ratios. Very interestingly, the halo stars with low [α/Fe] ratios tend to be on orbits biased to the outer halo. This could be the consequence of the slower formation of the outer halo relative to the inner one. In this situation, in fact, as can be inferred from Figure 6.3, chemical evolution models predict that in external galactic regions or external galaxies (e.g. the Magellanic Clouds), where the star formation has been slower than in the solar vicinity, we should expect lower [α/Fe] ratios than in the solar neighbourhood at the same [Fe/H].

Gratton et al. (1997a, 2000), as mentioned already in chapter 1, presented a compilation of data for [Fe/O] and suggested an even more complex picture for the formation of the Galaxy, showing that a net separation between the dissipational collapse model and the accretion scenario is rather artificial. It is more likely, in fact, that both processes were at work. In particular, they identified three kinematically distinct stellar populations: i) a population made of halo,

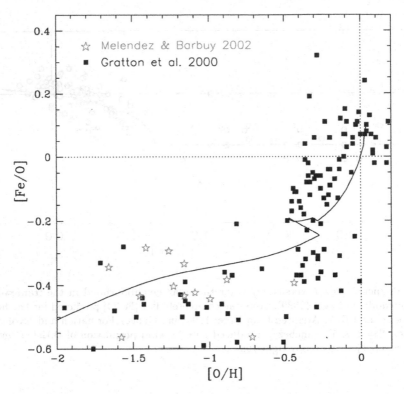

Figure 6.5. Predicted and observed [Fe/O] vs. [O/H] in the solar neighbourhood. Data from Gratton et al. (2000) and Melendez & Barbuy (2002). Model predictions from the two-infall model of Chiappini et al. (1997).

thick-disk and perhaps few bulge stars originating from an early fast dissipative collapse, ii) a population of thin-disk stars originating from a subsequent even more dissipative collapse which formed the thin-disk and, finally, iii) a population of thick-disk stars the origin of which should be different from the others, namely formed in satellite galaxies and then accreted by the Galaxy.

From Gratton et al.'s data there is also evidence (see Figure 6.5) for a long plateau where [Fe/O] is almost constant indicating a short timescale for the formation of the halo and part of the thick-disk stars (≤ 1 Gyr) through dissipative collapse. The accretion process which should be responsible for the thick-disk population should have acted also on a comparable timescale, although some later accretion of some low-mass fragments is possible. This scheme might naturally produce the discontinuity between thick- and thin-disk phases as observed in the data. In particular, what is observed in Figure 6.5 is that at [O/H]$\sim$ -0.25 dex the oxygen abundance is constant whereas the [Fe/O] ratio keeps increasing. It is worth noting that also [Fe/Mg] vs. [Mg/H] shows a similar behaviour as indicated by a study of Furhmann (1998) (see Figure 1.9). This can be interpreted, in the framework of the time-delay model, as a sudden decrease in the SFR, since in this case oxygen would not be much produced whereas Fe would continue to originate from the long-living systems giving rise to type Ia SNe. This sudden decrease in the SFR can perhaps be due to gas heating from SNe which might have triggered even a galactic wind (BS scenario). The duration of such period of low star formation, during which the gas was cooling, should be inside the range 0.5-3 Gyr, as estimated by chemical evolution models on the basis of the fact that this phase should have been larger than the typical timescale for type Ia SNe but not too long to create a clear discontinuity in the data. However, more data on more elements are necessary to better ascertain this point.

On the basis of the Gratton et al.'s data, Chiappini et al. (1997) developed a chemical evolution model aimed at reproducing mostly the halo and thin-disk star distributions, given the possible external origin of thick-disk stars. This model assumes two major gas infall episodes forming the halo together with part of the thick-disk and the thin-disk, respectively (see Section 6.6).

Another interesting approach to the chemical evolution of the Galaxy consists in a model for the chemical evolution of the solar neighbourhood region where the three components, halo, thick- and thin-disk are considered separately (Pardi et al. 1995). In this scheme, the three components evolve at different rates but they are connected through gas flows, in the sense that during the halo formation gas is lost to form the thick-disk which then accumulates with a time delay relative to the halo. The thin disk forms with an even longer time delay, from the residual halo and thick-disk gas (no accretion of extragalactic gas is considered), since a progressively longer timescale for the formation of the younger components is assumed. However, while the timescale for the formation of the thick-disk is similar to that of the halo, because the gas lost from the halo goes directly to form the thick-disk, the difference between the thin-disk and the halo is larger since the rate of formation of the thin-disk depends only on the gas lost from the thick-disk. Thus the formation of the thick-disk is an intermediate step which allows one to separate the halo from the thin-disk evolution. As a result of this, the thin disk forms out of already enriched material and on a timescale as long as 3-4 Gyr thus ensuring a fit to the G-dwarf metallicity distribution. In addition, this model predicts that the [O/Fe] vs. [Fe/H] relation is not unique but is given by the convolution of the [O/Fe] vs. [Fe/H] relations for each galactic component. This fact could help in explaining the spread in the observational data. In other words, this kind of dissipative collapse model assumes *a parallel formation*, as opposed to *a serial formation* as in the model of Matteucci and François (1989; 1992). In particular, no spread at low metallicities is predicted whereas for [Fe/H]> −1.0 dex a spread of the order of 0.3-0.4 dex is found (see Figure 6.8). The reason for this spread resides in the fact that the chemical evolution of the gas in the halo and in the thick-disk continues until the present time although the amount of star formation becomes negligible after 2 Gyr in the halo and after

8 Gyr in the thick-disk. Therefore, very few halo stars with [Fe/H]> −1.0 dex and very few thick-disk stars with [Fe/H]> −0.5 dex are predicted to exists, so that they cannot explain all of the observed spread, but probably only part of it. However, the main weakness of this model is represented by the failure in reproducing the thick-disk star metallicity distribution (see section 6.5). This is due to the fact that the assumption that the thick-disk formed out of the gas shed by the halo seems unrealistic.

To summarize what has been said in the previous sections, the collective evidence favours the view that the halo of the Galaxy formed by a combination of rapid, dissipative collapse (particularly in the innermost regions) and mergers with stellar systems such as dwarf spheroidals and dwarf irregulars (particularly in the outer regions). Concerning the formation of the other components, especially the thin-disk, the most reliable constraint resides in the distribution of the disk G-dwarfs as a function of metallicity. This distribution in fact, as we will see more in detail in the following sections, is strictly related to the mechanism of formation of the thin-disk at the solar circle, under the assumption of an initial mass function constant in time. Evidence for an *inside-out* formation of the thin-disk is supported by abundance gradients measured along the galactic disk. The existence of such gradients is, in fact, favored by the assumption that the innermost regions of the disk formed faster than the outermost ones (see Section 6.8).

6.4   SERIAL FORMATION

We will describe here in more detail an example of a model assuming a serial formation of the Galaxy. In such a model the Milky Way is assumed to form during a unique infall event where during the first 1 - 2 Gyr the halo form and during the next 3-4 Gyr there is the formation of the thick- and thin-disk. Since the star formation rate is assumed to depend on some power of the gas density, and the gas accumulates with a timescale τ which varies with the galactocentric distance, it emerges that the star formation rate reaches a maximum in correspondence to the maximum infall rate and then it declines steadily. This scenario was originally devised to study the chemical evolution of the disk (Chiosi, 1980) and in principle it is not correct to apply it to the halo phase as well. However, from the point of view of the chemical evolution, the *serial* models have proven able to reproduce the observed abundances in stars as well as other more realistic models such as the *parallel* model or the *two-infall* model.

The basic equations of this models resemble eq. (5.67) without the outflow term, $X_i W(t)$, and with the infall term defined as:

$$X_{iA}A(t) = X_{iA}a(r)e^{-t/\tau(r)} \tag{6.2}$$

where $a(r)$ is a parameter defined by reproducing the present time total surface mass density in the disk. In particular, to obtain $a(r)$ one sums over all the chemical elements and integrates equation (6.2) with respect to time in order to reproduce the present time (t_G) distribution of the total surface mass density along the disk:

$$\sigma(r, t_G) = a(r)(1 - e^{-t_G/\tau(r)})\tau(r) \tag{6.3}$$

for a fixed galactocentric distance r.

Then, $a(r)$ is obtained as follows:

$$a(r) = \frac{\sigma(r, t_G)}{\tau(r)(1 - e^{-t_G/\tau(r)})} \tag{6.4}$$

where $\sigma(r, t_G)$ represents the observed total surface mass density at the galactocentric distance r at the present time t_G.

The observed total surface mass density distribution can be approximated by an exponential law:

$$\sigma(r, t_G) = \sigma_D e^{-r/R_D} \tag{6.5}$$

where σ_D and R_D are two scale parameters.

The abundance X_{iA} represents the abundance of the element i in the infalling gas. Normally the abundances of the infalling gas are the primordial ones, namely with no metals.

The timescale for the complete formation of the disk $\tau(r)$ is a function of the galactocentric distance r and is defined to give a timescale at the solar neighbourhood of $\tau_\odot(8\text{-}10 \text{ kpc}) = 4$ Gyr. This particular timescale was imposed by reproducing the metallicity distribution of the G-dwarfs by Pagel and Patchett (1975). It is obviously a free parameter and it might change if the G-dwarf metallicity distribution changes, as is the case for the more recent distribution found by Rocha-Pinto and Maciel (1996) which shows a peak in the number of stars at a value of [Fe/H] different from that of Pagel and Patchett (1975) (see Figure 1.11).

It has been shown by Chiappini et al. (1997) that in order to reproduce the more recent G-dwarf metallicity distribution the timescale for the formation of the disk at the solar circle has to be longer than previously estimated, namely $\tau_\odot(8\text{-}10 \text{ kpc})=8$ Gyr (see Figure 5.4).

The dependence of τ on the galactocentric distance is also fixed by reproducing the abundance gradients in the galactic disk.

For example, Matteucci and François (1989) found that assuming the following relation:

$$\tau(r) = 0.464r - 1.59 \; Gyr \tag{6.6}$$

one can well reproduce the observed abundance gradients along the thin-disk (see Table 6.3 where model predictions are compared to the observations). The physical basis for the variation of τ with galactocentric distance resides in the results of the dynamical models of Larson (1976). The specific law for the growth of the timescale τ is instead an arbitrary choice (see also chapter 4). However, numerical experiments show that a quadratic or even an exponential law does not change significantly the model results.

The star formation rate is assumed to be a function of both the surface gas density, $\sigma_{gas}(r, t)$, and the total surface mass density $\sigma(r, t)$, in particular:

$$\psi(r, t) = \frac{d\sigma_{gas}(r, t)}{dt} = \nu[\frac{\sigma(r, t)\sigma_{gas}(r, t)}{\tilde{\sigma}^2(\tilde{r}, t)}]^{k-1}\sigma_{gas}(r, t) \qquad (6.7)$$

This formulation of the star formation rate was introduced by Talbot and Arnett (1975) under the assumption of a feed-back mechanism regulating star formation, as already discussed in chapter 3. The quantity $\tilde{\sigma}(\tilde{r}, t)$ represents the total surface mass density at a particular galactocentric distance $\tilde{r}$, originally introduced only for normalization purposes (see Chiosi, 1980).

The initial mass function, by mass, is assumed to be constant in space and time and to have the form:

$$\varphi(m) = Am^{-x_1} \qquad (6.8)$$

with $x_1 = 1.35$ for $m \leq 2M_\odot$ and:

$$\varphi(m) = Bm^{-x_2} \qquad (6.9)$$

with $x_2 = 1.7$ for $m > 2M_\odot$, and to be defined in the range $0.1 - 100M_\odot$. The two constants A and B are derived by assuming the normalization conditions:

$$\int_{0.1}^{2} Am^{-x_1}dm + \int_{2}^{100} Bm^{-x_2}dm = 1 \qquad (6.10)$$

and

$$A2^{x_1} = B2^{x_2} \qquad (6.11)$$

This kind of model predicts that the [α/Fe] ratios versus [Fe/H] evolve like the [O/Fe] shown in Figure 6.2. The initial plateau in the [O/Fe] for metallicities [Fe/H] < -1.0 dex is followed by a steady decline of the same ratio until the solar value is reached ([O/Fe]=0.0). In a self-consistent approach, it is better to normalize the predicted (O/Fe) ratio to the predicted solar (O/Fe)$_\odot$ rather than adopting the observed solar value. In fact, what matters is to reproduce the behaviour of the [O/Fe] vs. [Fe/H] rather than fitting perfectly the absolute abundances. Absolute abundances depend on all the main model assumptions and therefore do not represent a very good model constraint. Abundance ratios instead depend only on the yields, namely on the nucleosynthesis and the assumed IMF, and on stellar lifetimes, whereas the behaviour of the abundance ratios as functions of time and/or metallicity depends also on the star formation history. Therefore, if a chemical evolution model is able to reproduce the correct behaviour of the [O/Fe] vs. [Fe/H] this means that oxygen and iron have been assumed to be produced in the right proportions as well as that the star formation history is not far from the reality. In particular, the predicted behaviour shown in figure 6.2 tells us that at low metallicities the yields from massive stars predominate and the [O/Fe] ratio is larger than the solar value since most of the iron, produced by type Ia SNe, has not yet been restored into the ISM. It is worth noting that the O/Fe ratio produced by massive stars exhibits a dependence on the initial stellar mass, the particular dependence varying with the assumed yields. For example, in Figure 6.6 is shown the behaviour of the yields of Woosley (1987) weighted by the IMF as functions of the initial stellar mass. A close inspection of the figure shows that there is a general decrease of the O/Fe ratio with decreasing stellar mass. This decrease is reflected in the [O/Fe] ratio predicted for the ISM, which exhibits a plateau not perfectly flat. The more recent yields by WW95 (solar metallicity) and TNH95 are shown in Figure 6.7 where we report directly the O/Fe production ratios.

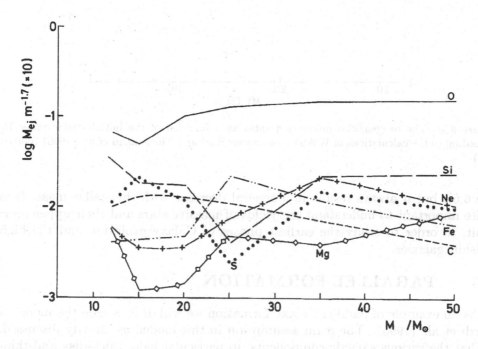

Figure 6.6. The masses ejected in the form of various elements as functions of the initial stellar mass as predicted by the yields of Woosley (1987) weighted by the Scalo (1986) IMF. From Matteucci and François 1989, M.N.R.A.S. Vol. 239, 885; reproduced here by kind permission of Blackwell Science LTD. (copy right 1989).

For the yields of Figure 6.7 the behaviour of the O/Fe produced by massive stars is even more complicated than in Woosley's previous calculations (Fig-

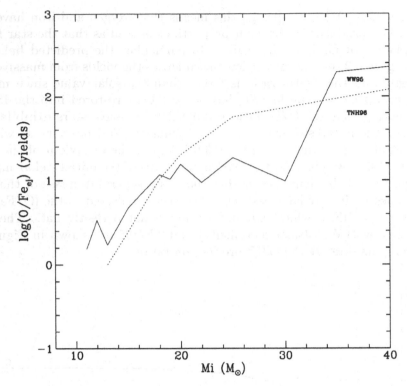

Figure 6.7. The oxygen/iron production ratio as a function of the initial stellar mass M_i according to the calculations of WW95 (continuous line) and Thielemann et al. (1996) (dotted line).

ure 6.6), but it always results in a general decrease with the stellar mass. It is quite important to understand the yields of massive stars and their upper mass limit, in order to study the earliest phases of Galaxy evolution and the high redshift galaxies.

6.5   PARALLEL FORMATION

As an example of parallel Galaxy formation we will discuss here the model of Pardi et al. (1995). The main assumption in this model, as already discussed, is that the various galactic components, in particular halo, thick-disk and thin-disk, start forming at the same instant but evolve at different rates. The three components are related to one another in the sense that the thick-disk forms out of the gas shed by the halo and the thin-disk out of the gas shed by the thick-disk.

In each zone the matter is assumed to be present in four forms : diffuse gas (g), molecular clouds (c), stars of both low (s_1) and high mass (s_2 for $M > 4M_\odot$)

and remnants (r). These quantities are all defined as the ratio between the mass of gas, clouds, stars and remnants relative to the total mass. Because of this subdivision in various phases for the gas and stars, the authors refer to their model as *multi-phase.*

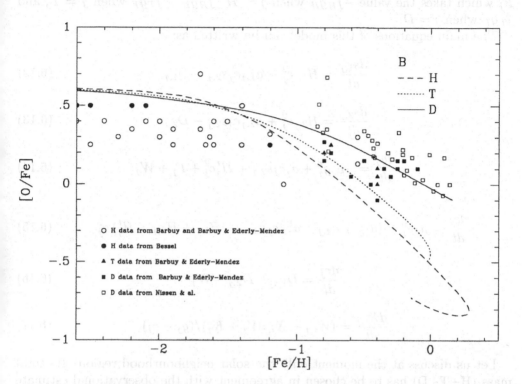

Figure 6.8. Predicted and observed [O/Fe] vs. [Fe/H] in the solar neighbourhood. The dashed line refers to the halo (H), the dotted line refers to the thick-disk (T) and the continuous line refers to the thin-disk (D). The label B refers to Model B of Pardi et al. (1995). From Pardi et al. 1995, Ap.J. Vol. 421, 491: reproduced here by kind permission of M.C. Pardi and the University of Chicago Press (copy right 1995).

The coupling between zones (halo thick-disk and thick-disk thin-disk) is obtained through the parameters f_H and f_T which represent the rates of gas accretion and in principle depend both on gravitational and dissipative processes but in reality they are two free parameters.

In the following model equations, the suffix j refers to each zone: $j = H$ for halo, T for thick-disk and D for thin-disk. The indexes 1 and 2 refer to low and high mass star ranges, respectively. The terms D_j and W_j are the rate of death and the rate of chemical element restoration, respectively. The terms H, K, a and μ represent the efficiencies of the various processes of star and cloud formation (see the original paper for details). In particular, the term K_j (see eq. 6.17) is 0 when $j = H$, is $f_{H}g_{H}(X_{iH} - X_{iK})$ where the X_i are the chemical abundances when $j = T$, and is $f_{K}g_{K}(X_{iK} - X_{iD})$ when $j = D$. The choice of $K_j = 0$ in the halo is because the authors assumed that there is no cloud formation during the halo phase. The coupling of zones is described by the term F_j which takes the value $-f_{H}g_{H}$ when $j = H$, $f_{H}g_{H} - f_{T}g_{T}$ when $j = T$, and $f_{T}g_{T}$ when $j = D$.

The main equations of this model can be written as:

$$\frac{ds_{1,j}}{dt} = H_{1,j}c_j^2 + a_{1,j}c_j s_{2,j} - D_{1,j} \tag{6.12}$$

$$\frac{ds_{2,j}}{dt} = H_{2,j}c_j^2 + a_{2,j}c_j s_{2,j} - D_{2,j} \tag{6.13}$$

$$\frac{dg_j}{dt} = -\mu_j g_j^n + a'_j c_j s_{2,j} + H'_j c_j^2 + F_j + W_j \tag{6.14}$$

$$\frac{dc_j}{dt} = \mu_j g_j^n - (a_{1,j} + a_{2,j} + a'_j) c_j s_{2,j} - (H_{1,j} + H_{2,j} + H'_j) c_j^2 \tag{6.15}$$

$$\frac{dr_j}{dt} = D_{1,j} + D_{2,j} - W_j \tag{6.16}$$

$$\frac{dX_{i,j}}{dt} = (W_{i,j} - X_{i,j}W_j + K_j)/(g_j + c_j) \tag{6.17}$$

Let us discuss at the moment only the solar neighbourhood region. Its total mass (H+T+D) has to be chosen in agreement with the observational estimate of the present local surface mass density. It is worth noticing that with this kind of model one can follow separately stars and gas belonging to each galactic component at the same time, whereas in the serial formation approach there is no overlapping in the formation of stars in each component. However, this model introduces more free parameters than the other and this represents always a complication and a problem for models of galactic chemical evolution and requires the definition of a larger number of observational constraints which should be honored by the model.

Figure 6.8 shows the predictions of this model for the behaviour of the [O/Fe] versus [Fe/H]. In this figure the predicted curves for the three galactic components are compared with observational data belonging to the same components. As already discussed, this model predicts a spread in the [O/Fe] ratio for a fixed [Fe/H] starting from [Fe/H] > -1.6 dex. The occurrence of the change in slope of the [O/Fe] ratio also occurs at different [Fe/H] values in the different galactic components. The existence of such a "knee" has been extensively discussed in the literature although there is still some confusion about the reason for its occurrence and therefore we will repeat here some concepts. The change in slope, in models adopting an IMF constant in time, is due to the appearance of the iron restored by type Ia SNe, which starts occurring at a precise time due to the assumed type Ia supernova progenitors and at a precise metallicity, [Fe/H], due to the assumed history of star formation. The type Ia supernova progenitors adopted in the models (serial and parallel) are the same as in eq. (5.67) and imply that the first type Ia SNe occurred not before several 10^7 years after the birth of the first stars but the bulk of the iron restoration takes up to 2 Gyr. In the parallel model we are discussing now, the star formation rate is different in the different galactic components. In particular, in the halo phase the star formation rate is quite high and we could describe the process as a starburst (see Figure 6.9). Under these conditions the gas in the halo is consumed very quickly and therefore oxygen production is inhibited at early times while some Fe continues to be produced by type Ia SNe. This is why the decline of the [O/Fe] ratio occurs at very low metallicities. The situation of the thick-disk is similar although less extreme. In fact, also in the thick-disk the star formation rate is high but not as much as in the halo. The situation changes in the thin-disk which is assumed to form slowly and therefore the predicted star formation rate is much lower than in the other two galactic components. As a result of that, the gas is consumed slowly and star formation is active until the present time. In this case, the knee of [O/Fe] occurs owing only to the restoration of the bulk of iron. This occurs at roughly [Fe/H]=-1.0 dex, in agreement with the observational data and corresponding to an age of about 2 Gyr, in marginal agreement with the serial model.

An interesting aspect of Figure 6.8 is that it predicts the possibility of having halo stars with solar O/Fe ratio. However, the predicted number of halo stars with solar metallicity is extremely small since the star formation rate in the halo goes practically to zero after about 2 Gyr from the beginning of star formation. The curves shown in Figure 6.8, in fact, refer to the abundance ratio predicted for the gas, and do not necessarily indicate that there should be stars with those metallicities. An interesting test to this would be to find stars with halo kinematics and solar abundances. Stellar surveys have indeed indicated that there is some overlapping between halo and thick-disk stars but only at the halo thick-disk metallicity transition (Beers and Sommer-Larsen, 1995; Nissen and Schuster 1997). Unfortunately, a limit of the parallel model, as already discussed, is the fact that it is unable to fit simultaneously the metallicity distribution of stars in the different galactic components, as shown in Figure 6.10. This is because of the assumed link between the formation of the halo, thick and thin-disk. As a consequence of this, we cannot avoid the conclusion that the different galactic components should have formed independently, in particular the halo and the disk. On the other hand, as indicated by the very similar

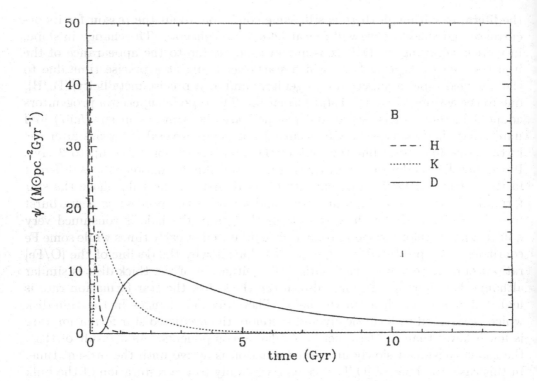

Figure 6.9. Predicted star formation rate in halo, thick- and thin-disk by the parallel model of Pardi et al. (1995) in the solar neighbourhood. The symbols are like in Figure 6.8. From Pardi et al. 1995, Ap.J. Vol. 421, 491: reproduced here by kind permission of M.C. Pardi and the University of Chicago Press (copy right 1995).

distribution of the specific angular momentum in bulge and halo stars (see figure 6.11), these two components should be linked as different parts of the same spheroid. This kinematical feature seems to be a robust indication that models linking the formation of the halo with that of the disk should be rejected.

6.6   TWO-INFALL FORMATION

An alternative approach to the serial and parallel Galaxy formation is to assume that different galactic structures (at least halo and disk) formed in a completely independent way by means of separate infall episode. This was proposed by Chiappini et al. (1997) and is based on the following idea: a first infall episode gave rise to the halo and may be part of the thick-disk and the bulge, this latter being formed by the gas accumulating at the centre of the spheroid, then a second infall episode, completely independent from the first one, formed the thin-disk . The timescale for the formation of the spheroid and part of the thick-

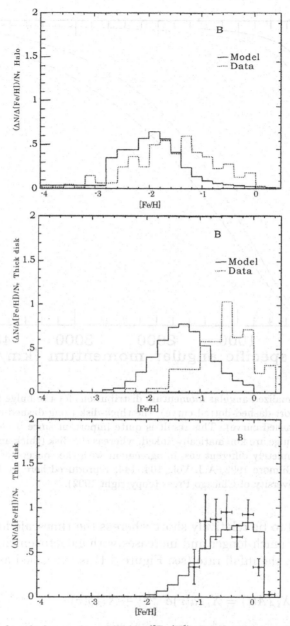

Figure 6.10. Predicted and observed metallicity ([Fe/H]) distribution in solar neighbourhood stars. The top panel shows the distribution of halo stars, the intermediate panel refers to the thick-disk stars and the bottom panel refers to the thin-disk stars. The data for halo and thick-disk stars are from Sandage and Fouts (1985), whereas those for the thin disk are from Sommer-Larsen (1991). From Pardi et al. 1995, Ap.J. Vol. 421, 491: reproduced here by kind permission of M.C. Pardi and the University of Chicago Press (copy right 1995).

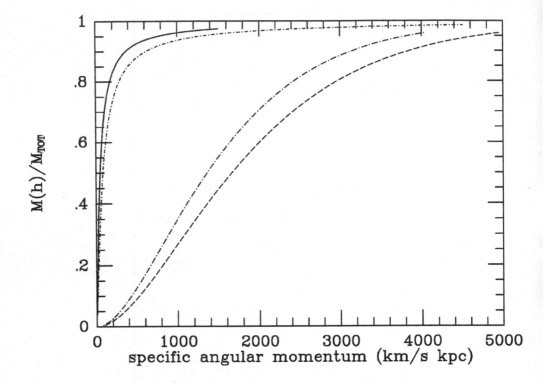

Figure 6.11. Normalized angular momentum distributions for the bulge (solid curve), the $r^{1/4}$ spheroid (halo, short-dashed-dotted curve), the thick-disk (long-dashed-dotted curve) and the thin-disk (long- dashed curve). This result is quite important since it clearly shows that only the halo and the bulge are kinematically linked, whereas the disk (thick and thin) seems to have formed out of completely different gas, in agreement with the concept of the two-infall model. From Wyse and Gilmore 1992, A.J. Vol. 104, 144; reproduced here by kind permission of R. Wyse and the University of Chicago Press (copy right 1992).

disk is assumed to be relatively short whereas the timescale for the formation of the thin-disk is much longer and increases with galactocentric distance.

In particular, the infall rate (see Figure 4.1) is expressed as;

$$X_{iA}A(t) = X_{iA}a(r)e^{-t/\tau_1} + X_{iA}b(r)e^{-(t-t_{max})/\tau_2(r)} \tag{6.18}$$

where X_{iA} are the primordial abundances for both infall episodes and $a(r)$ is obtained by reproducing the total surface mass density at the end of the halo thick-disk phase as in (6.4) and τ_1 is the timescale for the formation of the halo thick-disk and is assumed to be ~ 2 Gyr. The term $b(r)$ is again derived as in (6.4) by imposing to reproduce the present time total surface mass density in the thin disk. The quantity t_{max} is the time of the maximum infall onto the disk.

Finally, $\tau_2(r)$ is the timescale for the formation of the thin-disk and is a function of the galactocentric distance:

$$\tau_2(r) = 0.875r - 0.75 \ Gyr \tag{6.19}$$

The relation (6.19) is assumed on the basis of having $\tau_2(r_\odot) = 8$ Gyr, with $r_\odot = 10$ kpc. This long timescale for the formation of the solar neighbourhood clearly implies that the disk cannot have formed out of the gas lost by the halo. The star formation rate is the same as in equation (6.7) but there is a threshold in the gas density below which the SFR is assumed to go to zero, namely:

$$\psi(r,t) = \frac{d\sigma_{gas}(r,t)}{dt} = \nu[\frac{\sigma(r,t)\sigma_{gas}(r,t)}{\tilde{\sigma}^2(\tilde{r},t)}]^{1.4}\sigma_{gas}(r,t) \tag{6.20}$$

when $\sigma_{gas} > \sigma_{th}$ and

$$\psi(r,t) = 0 \tag{6.21}$$

when $\sigma_{gas} \leq \sigma_{th} = 7 M_\odot pc^{-2}$.

This value for the gas density threshold is in agreement with observational estimates (see chapter 3). The efficiency of star formation is also assumed to vary from the halo to the disk being higher in the former, as shown in Figure 6.12.

The main consequence of the two-infall model is that the evolution of the thin-disk is completely disentangled from the evolution of the inner halo thus allowing the reproduction of the completely different metallicity distributions in the different components (see in Figure 5.4 the very good fit of the G-dwarf metallicity distribution).

The second important consequence of this model is that the existence of a threshold in the star formation leads naturally to a halt in the star formation rate at the end of the first infall episode, thus creating the hiatus in the star formation suggested by the dynamical model of Larson (1976) and observed in the [Fe/O] vs. [O/H], shown in Figure 6.5, where the predictions of the two-infall model are also present and indicate a good agreement with the data. In Figure 6.12 the effect of adopting a threshold in the star formation rate, which induces an oscillatory behaviour, is evident; in particular, the SFR goes to zero every time the surface gas density goes below the threshold. This happens mostly at the end of the first phase and it contributes to keep a constant surface gas density for the last few Gigayears of the life of the disk, thus solving the problem of gas consumption on timescales shorter than the galactic age, as discussed in chapters 4 and 5.

6.7 THE EVOLUTION OF THE SOLAR NEIGHBOURHOOD

We summarize here the results concerning the temporal evolution of the abundances of the heavy elements in the solar vicinity region.

198

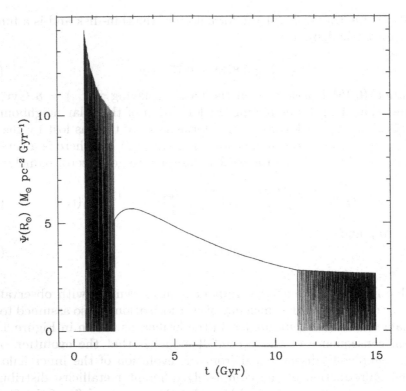

Figure 6.12. The star formation rate in the halo-(thick-disk) and in the thin disk, as predicted by the two-infall model of Chiappini et al. (1997). The oscillations are due to the adoption of a threshold gas density for the SFR as discussed in the text. From Chiappini et al. 1997, Ap.J. Vol. 477, 765; reproduced by kind permission of C. Chiappini and the University of Chicago Press (copy right 1997).

6.7.1 THE EVOLUTION OF THE ELEMENTS FROM CARBON TO ZINC

In Figures 6.13, 6.14 and 6.15 predictions are shown of the two-infall model concerning C, N, O, Mg, Ne, Si, S, Cu and Zn as functions of Fe. The model results refer to two different sets of yields for massive stars, namely those of Woosley and Weaver (1995) ($Z = Z_\odot$ their case B for masses larger than $30M_\odot$) and those of Thielemann et al. (1996). The data and the statistical analysis are the same as in Figures 1.1 and 1.2 in chapter 1.

Concerning carbon, the data best fit shows an almost constant, almost solar, [C/Fe] ratio over the whole [Fe/H] range. The most straightforward interpretation of this behaviour in the framework of the time-delay model is that both C and Fe are produced on the same timescales, namely they are mostly produced by stars with low and intermediate masses and partly by massive stars. Therefore, the comparison between data and models tends to support the idea that C is mostly supplied by low and intermediate mass stars. However, the plot of [C/O] versus [Fe/H] (see Figure 6.16) shows a steep increase of the [C/O] ratio with increasing metallicity and the same models of Figures 6.13-6.15 do not reproduce such an increase unless suppressing completely the primary nitrogen produced in low and intermediate mass stars (this can be achieved by suppressing the third dredge-up phase and the result is shown in Figures 6.16 and 6.17). On the other hand, the yields of M92 assuming a quite strong mass loss (see Figures 2.17 and 2.18) underproduce O in massive stars of metallicity higher than solar and overproduce C with respect to the standard yields without mass loss. Figure 6.17 shows that the high [C/O] ratio at high [Fe/H] is well reproduced by the yields of M92. However, these yields refer only to massive stars (9-$120M_\odot$) and do not consider the heavier elements such as Si and Fe, therefore it is not possible to compute self-consistent models of galactic evolution predicting [C/O] and [Fe/H] at the same time.

The best fit to the [N/Fe] ratio indicates instead an almost continuous decrease of this ratio with decreasing [Fe/H], with a flattening for [Fe/H] < -2.0 dex; but it is not clear if this effect is real or if it is due to the large spread in the data at these low metallicities. The models do not reproduce the initial flattening since they assume that N is produced as a secondary element in massive stars. This fact may suggest that a fraction (or all) of N in massive stars should have a primary origin. On the other hand, for higher metallicities the model reproduces well the observational trend, namely when the contribution from intermediate and low mass stars appears. The N produced in these latter is partly secondary and partly primary (Renzini and Voli's yields) but the main effects of the growth of [N/Fe] are the progressively longer timescales of N progenitors.

Therefore, the origin of C and N is still a matter of debate and we need more precise abundance measurements at low metallicities together with homogeneous sets of nucleosynthesis calculations all over the relevant range of stellar masses (1-$120M_\odot$). An example of a homogeneous set of calculations, at least concerning C and N, can be found in Portinari et al. (1998).

The α-elements such as Mg, Si and Ca seem to be very well reproduced by both sets of yields. Concerning S, the data are still very limited and are missing for the halo stars. Oxygen data show a slight increase of the [O/Fe] with decreasing [Fe/H]. Notice that the data from Israelian et al. (1998), which show a stronger increase (see chapter 1), are not included. The slight increase of [O/Fe] at low metallicity is predicted also by the model with both sets of yields and this is due to the fact that the y_O/y_{Fe} yield ratio increases with increasing stellar mass, as shown already in Figure 6.7. The yields of TNH95 give higher [α/Fe] ratios at low metallicities than the yields of WW95 and this is mainly because they predict less iron (see chapter 2). Unfortunately, it is not possible to distinguish clearly among them, given the uncertainties still present in the data at very low metallicities. However, if the steep increase in the [O/Fe] ratio were real, we should expect an even more important role of SN Ia as Fe producers. Numerical experiments, in fact, show that if one suppresses type II SNe as Fe producers the predicted [O/Fe] decreases steadily with increasing metallicity. On the other hand, they show that if one suppresses type Ia SNe as Fe producers, in the framework of the standard nucleosynthesis, the predicted [O/Fe] remains almost constant over the whole [Fe/H] range.

The evolution of copper and zinc (elements beyond the iron peak) shown in Figure 6.15 is reasonably reproduced by the model. The main assumption is that most of Zn and Cu come, like iron, from type Ia SNe. These elements are also partly produced in massive stars as s-process elements, namely as a consequence of slow neutron capture on Fe-seed nuclei mainly during core He-burning. The assumption that Zn evolves in lockstep with Fe is particularly interesting, since Zn is the most commonly measured element in high redshift objects such as DLA, owing to the fact that it is almost unaffected by dust depletion, and therefore can be used as a substitute for Fe in the plots of abundance ratios versus [Fe/H].

6.7.2 THE EVOLUTION OF S- AND R-PROCESS ELEMENTS

We discuss here the evolution of a typical s- and a typical r-process element, namely barium and europium, respectively. As already discussed in chapter 2, s-process elements should behave like secondary elements since they are produced in proportion to the initial Fe abundance. Barium in particular should be mostly produced in low mass AGB stars as an s-process element and therefore restored into the ISM with a delay relative to elements mainly produced in massive stars, such as europium which is supposed to be mainly an r-process element. However, the uncertainties in the nucleosynthesis of these elements require a comparison between the predictions of chemical evolution models and abundance data in order to constrain the nucleosynthesis mechanisms. In Figures 6.18 and 6.19 predictions are shown of the parallel model of chemical evolution, described before, concerning [Ba/Fe] and [Eu/Fe] versus [Fe/H]. The authors (Travaglio et al. 1999a) conclude that the evolution of Ba cannot be explained by assuming that Ba is only an s-process element mainly produced in stars of 2-4 $M_\odot$, but an r-process origin for it should also be assumed. In fact, in the former hypothesis, a very late appearance of Ba is expected owing to its long-life progenitors and to its dependence on metallicity (Fe abundance), at variance with the observational data showing that barium must have appeared already between [Fe/H] = -4.0 dex and -3.0 dex. However, in spite of the good fit of the general trend of [Ba/Fe] versus [Fe/H], a good explanation for the observed spread at low metallicities is still missing. The general interpretation of such a spread, which is considered to be real, is the inhomogeneity in the halo gas which is, of course, a quite plausible explanation. However, before drawing conclusions on such a spread occurring

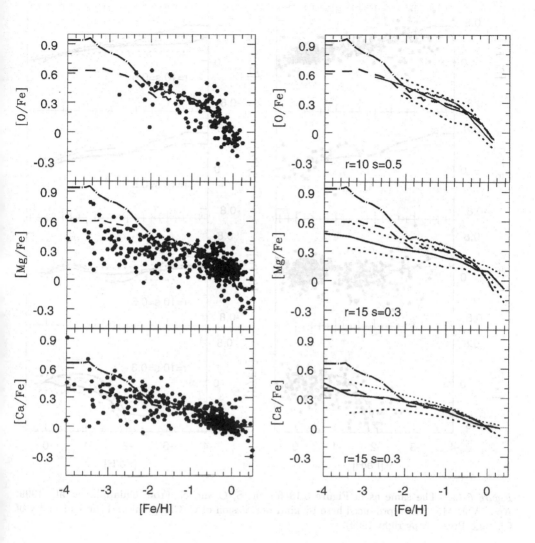

Figure 6.13. Abundance ratios [(O, Mg, Ca)/Fe] as functions of [Fe/H]. In the left panels the model predictions are compared to the data which are the same as in Figure 1.2; the dashed lines refer to the predictions by the two-infall model with the yields of Woosley and Weaver (1995), whereas the dashed-dotted lines refer to the yields of Thielemann et al. (1996). In the right panels are shown the summary lines compared with the model predictions; the parameter r represents the number of data points (10 or 15) in the estimation window and s refers the amount of smoothing (the fraction of data included at each location) in the summary lines (see also description of Figure 1.2). From Chiappini et al. 1999, Ap.J. Vol. 515, 226; reproduced here by kind permission of C. Chiappini and the University of Chicago Press (copy right 1999).

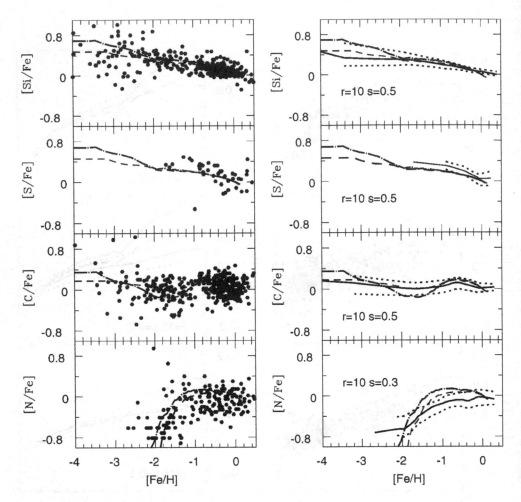

Figure 6.14. The same as in Figure 6.13 for Si, S, C and N. From Chiappini et al. 1999, Ap.J. Vol. 515, 226; reproduced here by kind permission of C. Chiappini and the University of Chicago Press (copy right 1999).

mostly in s-process elements, such as Sr and Ba, but not as much in α-elements, we should wait for more observational data. In fact, Ryan (2000) in a very recent analysis has shown that the spread in Ba is much smaller than the spread in Sr, and this may suggest that the origin of the spread is not yet clear.

On the other hand, europium should be mainly produced by rapid neutron capture during the SN II explosions. In particular, low mass SN II ($8.0 \leq M/M_\odot \leq 10.0$) should be the main contributors to this element. The data, in fact, show that Eu should appear after Fe in order to explain the data in the range $-3.0 \leq [Fe/H] \leq -2.0$, where the $[Eu/Fe] \leq 0$. In the framework

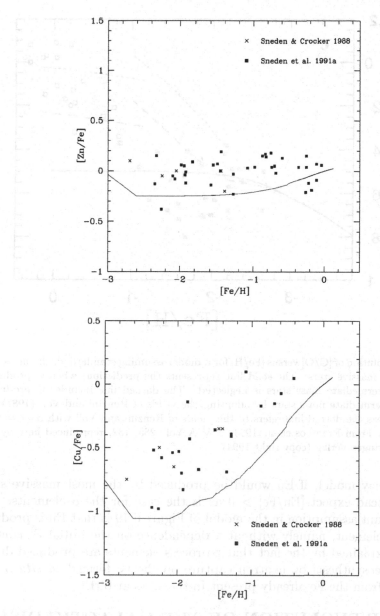

Figure 6.15.   The same as Figure 6.14 for Cu and Zn. The model predictions (best model of Chiappini et al. 1997) are indicated by the continuous lines. The nucleosynthesis prescriptions adopted for Cu and Zn assume that they are produced both during quiescent He-core burning in massive stars as s-process elements (secondary production) and in explosive nucleosynthesis in type Ia SNe (primary production). Detailed nucleosynthesis prescriptions can be found in Matteucci et al. (1993). From Chiappini et al. 1997, Ap.J. Vol. 477, 765; reproduced here by kind permission of C. Chiappini and the University of Chicago Press (copy right 1997).

204

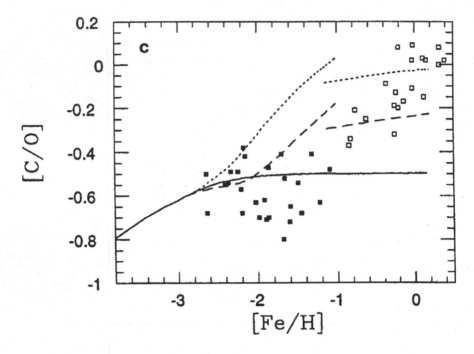

Figure 6.16. Evolution of [C/O] versus [Fe/H] for a model assuming standard yields (metallicity independent) for massive stars. The solid line represents the predictions when C production from low and intermediate mass stars is neglected. The dashed line includes C production from low and intermediate mass stars by adopting the yields of Renzini and Voli (1981) with $\alpha_{ML} = 1.5$ whereas the dotted line refers to the yields of Renzini and Voli with $\alpha_{ML} = 0$ (no third dredge-up). From Prantzos et al. 1994, A & A Vol. 285, 132; reproduced here by kind permission of Springer Verlag (copy right 1994).

of the time-delay model, if Eu would be produced by the most massive stars, we should instead expect [Eu/Fe] > 0 as is the case for the α-elements. The second important assumption in the model of Figure 6.19 is that Eu is produced as a primary element, namely without a dependence on the initial Fe content. This can be explained by the fact that r-process elements are produced during explosive nucleosynthesis by neutron capture on the Fe formed *in situ* by the stars and not from the Fe already present in the stars at birth.

6.8 THE EVOLUTION OF THE GALACTIC DISK
6.8.1 ABUNDANCE GRADIENTS

The existence of abundance gradients along the galactic disk seems now to be well established as suggested by different data sources, already mentioned in Sections 1.8 and 1.9 and summarized in Table 1.3.

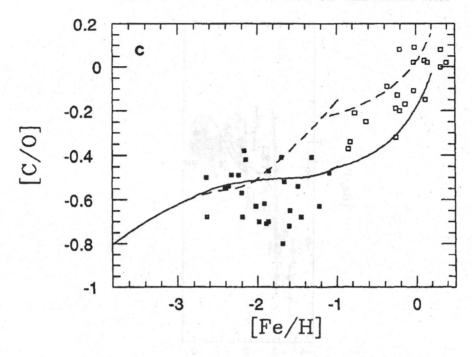

Figure 6.17. Evolution of [C/O] versus [Fe/H] for a model assuming M92 yields for massive stars. The solid line represents the predictions when C production from low and intermediate mass stars is neglected. The dashed line includes C production from low and intermediate mass stars by adopting the yields of Renzini and Voli (1981) with $\alpha_{ML} = 1.5$. From Prantzos et al. 1994, A & A Vol. 285, 132; reproduced here by kind permission of Springer Verlag (copy right 1994).

We recall that these sources are:

- *HII regions*: observations of abundance gradients from galactic HII regions indicate that O/H shows a gradient of the order of $\frac{\Delta log(O/H)}{\Delta R_G} = -0.07$ dex kpc^{-1} between 4 and 14 kpc along the galactic disk. The gradient of N seems to be slightly steeper than the one of oxygen. A flattening of O/H, N/H and S/H gradients for $R_G > 12$ kpc has been also suggested but there is no general agreement on this point.

- *Planetary Nebulae*: O/H, Si/H, Ne/H, Ar/H gradients from PNe are very similar to those found from HII regions, i.e. they are between $-0.04 - -0.07$ dex kpc^{-1} in the galactocentric distance range 4 - 14 kpc.

- *B type stars*: recent studies have found a gradient of oxygen of -0.07 ± 0.01 dex kpc^{-1} between 6 and 18 kpc and -0.07 ± 0.02 dex kpc^{-1} between 5 and 14 kpc, respectively. These results agree with the gradient estimated from HII regions and with the planetary nebula data.

206

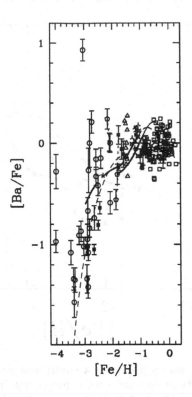

Figure 6.18. Galactic evolution of [Ba/Fe], where part of Barium is formed as an r-process in low mass SN II and the rest as s-process in low mass stars. The galactic model is that of Travaglio et al. (1999a). From Travaglio et al. (1999b), in "Nuclei in the Cosmos V", ed. N. Prantzos p. 531; reproduced here by kind permission of C. Travaglio and the University of Chicago Press (copy right 1999).

- *Open clusters*: the iron abundance measured recently in open clusters shows a gradient for [Fe/H] of $\frac{\Delta[Fe/H]}{\Delta R_G} = -0.06 \pm 0.01$ dex kpc^{-1} between 7 and 16 kpc (see chapter 1).

It is therefore encouraging that different data sources seem to agree at least on the value of the oxygen gradient. However, from a theoretical point of view we expect that the gradients of all the other heavy elements (C, N, Fe, Ar. etc..) should not be exactly the same as oxygen because of the differences due to the different nucleosynthetic production mechanisms and stellar progenitors of each element.

The question we should ask now is : *how can we theoretically obtain abundance gradients along the galactic disk?*

Several mechanisms have been proposed in the last years, although most of them ultimately consist in postulating a gradient in the SFR:

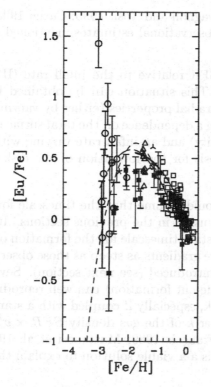

Figure 6.19. Galactic evolution of [Eu/Fe], where most of Eu is assumed to originate from low mass SN II but starting from [Fe/H] $\leq$ -1.5 the s-process component of Eu starts to appear. The model is from Travaglio et al. (1999a). From Travaglio et al. (1999b), in "Nuclei in the Cosmos V", ed. N. Prantzos p. 531; reproduced here by kind permission of C. travaglio and the University of Chicago Press (copy right 1999).

- i) variations in the gas fraction as a function of the galactocentric distance R_G (original idea of the Simple Model) but the predicted gradients, obtained by applying the solution of the Simple Model, are too flat when compared to observations.

- ii) Variations in the IMF but most of the suggested variations (massive stars increase with decreasing metallicity Z) tend to destroy the gradient! Carigi (1996) tried the opposite case, where low mass stars increase with decreasing metallicity, and showed that steep abundance gradients can be obtained but that the same model does not reproduce other observational constraints in the solar neighbourhood.

- iii) Metal dependent stellar yields (i.e. less metals are produced at higher Z, M92), produce shallow gradients along the galactic disk, as it has been shown by several papers (Giovagnoli and Tosi, 1995; Carigi 1996), not in agreement with the most recent observational estimates mentioned above.

- iv) Variations of the SFR relative to the infall rate (IR), namely SFR/IR, as a function of R_G. This situation can be obtained by assuming a SFR strongly depending on radial properties (either by varying the star formation efficiency or introducing a dependence on the total surface mass density as well as the surface gas density) and an infall rate varying with R_G. In particular, the *inside-out* hypothesis for the formation of the disk creates exactly this situation.

In other words, one should assume that the timescale for disk formation is a function of R_G, as discussed in the previous sections. It can be shown that models assuming a constant timescale for the formation of the disk are unable to reproduce abundance gradients as steep as those observed, unless some additional hypothesis is introduced (see next section). Several authors showed that biased infall (inside-out formation) can well reproduce steep abundance gradients along the disk, especially if coupled with a star formation rate proportional to some power k of the gas density $SFR \propto \sigma_{gas}^k$ with $k > 1$ (Tosi 1988; Matteucci and François 1989; Matteucci et al. 1989; Chiappini et al. 1997). Therefore, this is a a viable solution to explain the existence of abundance gradients .

- vi) Radial flows along the galactic disk are not sufficient by themselves to produce steep gradients but they are effective if coupled with biased infall or a threshold in the SFR under very specific conditions (see chapter 4).

- vii) A biased outflow assuming a stronger and earlier wind in the external regions than in the internal ones can also produce steep abundance gradients in the old stars, as shown for elliptical galaxies by Martinelli et al. (1998). However, this hypothesis is unlikely for the galactic disk and disks in general, since it implies a halt in the star formation after the galactic wind occurring first in the outermost disk regions, whereas star formation in disks is an ongoing process. In addition, the existence of continuous outflows in galactic disks is unlikely, because it would wash out the abundance gradients (see chapter 5) and would alter the $[\alpha/Fe]$ vs. $[Fe/H]$ relation in the solar neighbourhood, as we will see in the next sections. It is worth noting that biased infall and biased outflow would predict an opposite trend for the $[\alpha/Fe]$ ratios as functions of galactocentric distance in the framework of the time-delay model. In fact, the biased infall predicts that these ratios tend to diminish with the galactocentric distance whereas the biased outflow predicts the contrary.

6.8.2   ABUNDANCE GRADIENTS AS PREDICTED BY MODELS

Tosi (1996) analyzed the results of "best" models for the evolution of the galactic disk by several different authors and compared their results with observations. These models were aimed at reproducing observational results concerning the abundances of D, ^3He, C, N and O. What all of the examined models have in common is that they all take into account the finite stellar lifetimes, thus avoiding the instantaneous recycling approximation (I.R.A.), and they all do not explicitly include dynamics. The major assumptions made by these authors are summarized in Table 6.1, where all the times are expressed in Gyr and all the distances in kpc. Most of the authors assumed an age for the galactic disk of $T = 13$ Gyr and a solar galactocentric distance of $R_\odot \sim 8$ kpc.

Table 6.1.   Input Parameters of the Best Models

Author[a]	T	$R_\odot$	IMF[b]	SFR	θ	yields[c]
Carigi	13	8	Kr93	$A\Sigma_g^{1.4}(R)\Sigma_{tot}^{0.4}(R)$	0.6R-1.8	RV81+M92
Ferrini	13	8	Fe90	multiphase	1-2	RV81+WW86
Matteucci	13	10	Sc86	$A\Sigma_g^{1.1}(R)\Sigma_{tot}^{0.1}(R)$	0.464R-1.59	RV81+WW94
Prantzos-a	13.5	8.5	Kr93	$0.3\Sigma_g(R)/(R/R_\odot)$	3	M95+RV81+WW93
Timmes	15	8.5	Salp	$A\Sigma_g^2(R)$	4	RV81+WW94
Tosi-1	13	8	Ti80	$A(R)\exp(-t/\tau)$ $\tau = 15$	∞	RV81+CC79

a) References for these models are: Carigi 1994 and 1996; Pardi & Ferrini 1994, Ferrini et al. 1994, and Galli et al. 1995; Matteucci and François 1989; Prantzos & Aubert 1995, Prantzos 1996, and Prantzos et al. 1996; Timmes et al. 1995; Tosi 1988, Giovagnoli & Tosi 1995, and Dearborn et al. 1996.

b) IMF references are: Kroupa et al. 1993, Ferrini et al. 1990, Scalo 1986, Salpeter 1955 and Tinsley 1980.

c) References for the adopted yields are: Marigo et al. 1996 and Renzini & Voli 1981 for low and intermediate mass stars, Maeder 1992, Woosley & Weaver 1986, Weaver & Woosley 1993, Woosley & Weaver 1994, and Chiosi & Caimmi 1979 for massive stars. For Fe, Matteucci & Greggio's 1986 prescriptions are usually adopted.

The second column of Table 6.1 lists the adopted disk lifetimes, the third column the solar radius, the fourth column the adopted IMF, the fifth column the adopted law for the SFR. The sixth column lists the adopted infall e-folding times since all the models assumed that the disk formed by accretion of extragalactic (i.e. metal poor) material; the e-folding times of the models range from a minimum of 1-3 Gyr to a maximum of infinity (constant infall rate). The different e-folding times simulate different dynamical conditions, i.e. short times imply that the disk accretes gas only from the halo collapse whereas long times imply that the disk forms not only from the gas shed by the halo, but mostly from the intergalactic medium, for instance through the Magellanic Stream which is an on-going observed phenomenon. Some of the authors adopted a constant timescale for the formation of the disk at all radii whereas others assumed an inside-out disk formation. Finally, in the seventh column are reported the adopted prescriptions for stellar yields. Table 6.2 shows some relevant quantities predicted from the models described in Table 6.1.

Table 6.2. Output Parameters of the *Best* Models.

Author	SFR($R_\odot, t_\odot$)	SFR($R_\odot$,T)	g/m($R_\odot, t_\odot$)	g/m($R_\odot$,T)	infall(T)	$\Delta Y/\Delta Z$
Carigi	5.8	2.8	0.27	0.15	0.24	2.5
Ferrini	2.2	1.0	0.28	0.18	0.15	2.4
Matteucci	2.9	1.3	0.16	0.08	1.00	1.6
Prantzos-a	6.8	4.0	0.31	0.18	0.28	-
Timmes	?	4.1	?	0.10	0.23	4
Tosi-1	10.1	7.5	0.18	0.06	1.81	3

The predicted SFRs in the solar neighbourhood are in units of $M_\odot$ pc^{-2} Gyr^{-1} and refer to the time of the formation of the solar system (namely 4.5 Gyr ago) and to the present time (see columns 2 and 3, respectively). In columns 4 and 5 there are the ratios between the gas (g) and the total mass (m) at the time of solar formation and now, respectively. Column 6 lists the predicted infall rates at the present time (in units of $M_\odot yr^{-1}$) for the whole disk. Finally, the last column gives the predicted enrichment of He relative to metals over the galactic lifetime. In Figure 6.20 we report the present time O abundance gradient along the galactic disk predicted by all these models.

It is evident from Figure 6.20 that most of the chemical evolution models predict absolute oxygen abundances larger than the observed ones. This is due to the fact that most of the models are tuned to reproduce the solar abundances which are higher than the abundances observed in nearby HII regions such as Orion. This is a well known problem which could be perhaps solved by assuming that the Sun was born at a radius internal to the present one (orbital diffusion).

Concerning the predicted slopes of the O gradient, Figure 6.20 shows that the overall theoretical trends do not differ much from one another although some models show steeper gradients than others. This is mostly due to the different distribution of the SFR as a function of the galactocentric distance. In fact, abundance gradients in absence of radial flows, depend mostly on the history of star formation at different radii.

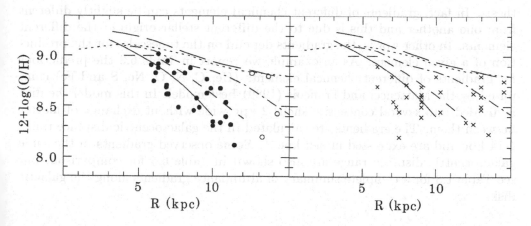

Figure 6.20. Radial distribution of the oxygen abundance at the present epoch as predicted by Table 6.1 models and derived by observations of young objects (HII regions in the left panel and B stars in the right panel). Models: *dash- dotted line*, Carigi; *short-dashed line* Ferrini; *long-dashed line* Matteucci; *dotted line* Prantzos; *solid-line* Tosi. From Tosi 1996, in "From Stars to Galaxies: the Impact of Stellar Physics on Galaxy Evolution", ASP Conf. Series, ed. C. Leitherer et al. p. 299; reproduced here by kind permission of M. Tosi and the Astronomical Society of the Pacific (copy right 1996).

In Figure 6.21 we show the radial distribution of oxygen in the galactic disk 3 Gyr ago, predicted by the models under discussion. This is to show the evolution of the gradients in time; the model predictions are compared with data on planetary nebulae. As one can see in this figure, different models predict different histories for the evolution of abundance gradients. Some of the models predict a steepening of the gradient in time whereas others predict a flattening. In general, models assuming an inside-out formation of the disk with long infall timescales in the outermost regions predict a steepening of the gradient in time. On the other hand, models where the infall is negligible at late times tend to flatten. Unfortunately, the observed data do not allow us to distinguish among the two possibilities, because:

- planetary nebulae seem to suggest that gradients steepen with time (Maciel and Köppen 1994); in fact, type III PNe (the oldest) have a flatter gradient than type II and I PNe, but dynamical effects could perhaps be responsible for this behaviour.

- Open clusters suggest no clear variation of gradients with time, and may be a flattening (see also chapter 1), but again dynamical effects could have played a role.

Before closing this section we discuss the fact that abundance gradients reflect not only the mechanism of formation of the disk but also the stellar nucleosynthesis. In fact, gradients of different chemical elements can be slightly different from one another and this is due to the different stellar origin of the different elements. In other words, the gradients depend on the timescales for the production of a given element. As an example we report in Table 6.3 the predictions for gradients of different chemical elements (He, C, N, O, Ne, S and Fe) computed by the Matteucci and François (1989) best model. In this model the disk is divided into several concentric shells 2 kpc wide without exchange of matter between them. The gradients are calculated in the galactocentric distance range 4-14 kpc and are expressed in dex kpc^{-1}. Some observed gradients in the same galactocentric distance range are also shown in Table 6.3 for comparison, but see Table 1.3 for a complete summary of abundance gradients along the galactic disk.

Table 6.3. Predicted and observed abundance gradients in the range 4–14 kpc

Element	Gradient ($\frac{\Delta log(X/H)}{\Delta R_G}$)	Observations
He	-0.0085	-0.01 ±0.008[1]; -0.078 ±0.023[4]
N	-0.0850	-0.09 ±0.015[1]
^{12}C	-0.0660	-0.035 ±0.014[4]
^{13}C	-0.0750	
O	-0.0650	-0.07 ±0.02[1]
Ne	-0.0480	-0.056 ±0.007[5]
Mg	-0.0500	-0.082 ±0.026[4]
Si	-0.0580	-0.107 ±0.028[4]
S	-0.0500	-0.01 ±0.02[1]; -0.077 ±0.011[5]
Fe	-0.0700	-0.09 ±0.017[2]; -0.06 ±0.01[3]

1) Shaver et al. (1983)
2) Friel and Janes (1993)
3) Friel (1999)
4) Gummersbach et al. (1998)
5) Maciel & Köppen (1994)

From the gradients in Table 6.3 we notice that mostly from a theoretical point of view the different chemical elements show slightly different gradients: for example, the gradient of oxygen is slightly flatter than the gradients of N and Fe. The reason for this can be understood in the following way: oxygen is produced by massive stars on short timescales, of the order of million years whereas N and Fe are mostly produced by low and intermediate mass stars (N from single stars and Fe from binaries, but on the same timescales). In addition, part of N is produced as a secondary product, namely proportional to the original oxygen abundance. The long timescales of production together with the secondary nature tend to increase the predicted gradient since they enhance the difference in the production of N between the innermost and the outermost regions of the galactic disk. This is due to the fact that the disk is assumed to form inside-out with a SFR much smaller in the outermost regions. The same happens to Fe although for this element only the timescale effect occurs, since Fe is a primary element. Therefore, we should expect a small gradient in abundance ratios such as N/O and α/Fe. Unfortunately, the precision of the available data does not yet allow such a detailed comparison, although the observed gradient of N seems slightly steeper than that of oxygen. The gradients of isotopic abundance ratios are also interesting because they can shed some light on the nucleosynthesis and stellar progenitors of the isotopes.

In Figure 6.22 we show the predictions for the radial gradient of $^{12}C/^{13}C$ as computed by some of the authors discussed above.

In general, Tosi concluded that the evolution of the C and O stable isotopes predicted by the various models indicate that something should be revised in stellar nucleosynthesis. In fact, the temporal and spatial behaviour of the $^{12}C/^{13}C$ ratio (as well as of $^{16}O/^{18}O$) predicted by adopting standard nucleosynthesis assumptions are inconsistent with the corresponding data (see Figure 6.22). A steeper gradient for $^{12}C/^{13}C$ is required to fit the observations; this can be achieved, for example, by assuming that novae are important producers of ^{13}C restored into the ISM with a large time delay, as suggested by the nature of such objects (white dwarfs in binary systems).

On the other hand, the correct behaviour of $^{12}C/^{13}C$ can be also achieved by adopting new C yields for low mass stars including deep extra mixing during the red giant phase associated with *cool bottom processing*, the net effect being that of producing more ^{13}C at late times. This mechanism (see for example Forestini and Charbonnel 1997; Sackmann and Boothroyd 1999) is also responsible for the ^{7}Li production in low mass red giants and for destroying ^{3}He in these same stars, thus alleviating the problem of the overproduction of this element by standard nucleosynthesis (see section 6.9).

For the O isotopes the problem resides in the fact that the predicted $^{16}O/^{18}O$ and $^{17}O/^{18}O$ ratios in the ISM are respectively higher and lower by a factor 1.6 than those inferred from molecular cloud observations. Tosi (1982) suggested that the problem could be solved if the amount of ^{18}O ejected by stars could decrease after the epoch of formation of the solar system. From the nucleosynthetic point of view ^{18}O is a neutron-rich element, namely an s-process element which should show, as do all the s-process elements, a sort of secondary or even tertiary behaviour (see chapter 2). Alternative ways out claim that either the observational $^{16}O/^{18}O$ is overestimated by a factor of 1.6 or the solar system ratio is not representative of that of the local ISM 4.5 Gyr ago.

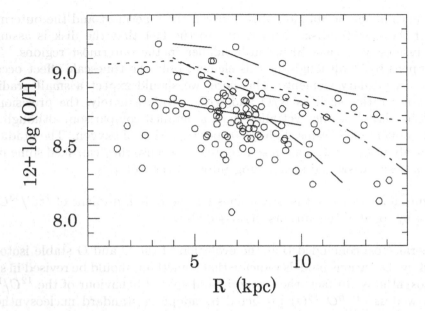

Figure 6.21. Radial distribution of the O abundance 3 Gyr ago as predicted by Table 6.1 models compared to observations of PNe II from Pasquali and Perinotto (1993). From Tosi 1996, in "From Stars to Galaxies: the Impact of Stellar Physics on Galaxy Evolution", ASP Conf. Series, ed. C. Leitherer et al. p.299; reproduced here by kind permission of M.Tosi and the Astronomical Society of the Pacific (copy right 1996).

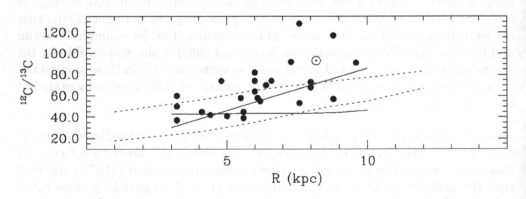

Figure 6.22. Radial distribution of the $^{12}C/^{13}C$ at the present epoch as derived from molecular cloud observations (Henkel et al. 1985) and as predicted by Prantzos (dotted line) and Tosi (solid) models. From Tosi 1996, in "From Stars to Galaxies: the Impact of Stellar Physics on Galaxy Evolution", ASP Conf. Series, ed. C. Leitherer et al. p.299; reproduced here by kind permission of M. Tosi and the Astronomical Society of the Pacific (copy right 1996).

6.8.3 RADIAL GAS AND STAR FORMATION DISTRIBUTION

In Figure 6.23 we report the predictions for the surface mass density distribution and the SFR distribution along the disk.

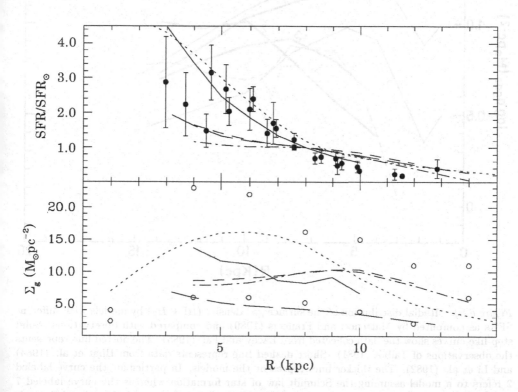

Figure 6.23. Radial distribution of the SFR (normalized to the value at the solar radius, top panel) and gas density (bottom panel) predicted for the present epoch by Table 6.1 models. Line symbols are as in Fig. 6.20. Dots and error bars in the top panel correspond to observational data on recent SFR as presented by Lacey and Fall (1985); the open circles in the bottom panel delimit the range of observed gas densities adopted by various modellers. From Tosi 1996, in "From Stars to Galaxies: the Impact of Stellar Physics on Galaxy Evolution", ASP Conf. Series, ed. C. Leitherer et al. p.299; reproduced here by kind permission of M.Tosi and the Astronomical Society of the Pacific (copy right 1996).

The predicted radial distribution of gas density and SFR are intimately related, especially in models assuming a dependence of the SFR on the surface gas density. In fact, the models with the flatter gas distribution show also the flatter SFR profiles. The only models in good agreement with both distributions are those of Prantzos and Tosi, and this is due to the fact that these authors adopt SFRs strongly declining outwards. The same models would predict also a steep abundance gradient. However, as shown by Matteucci and François (1989) the

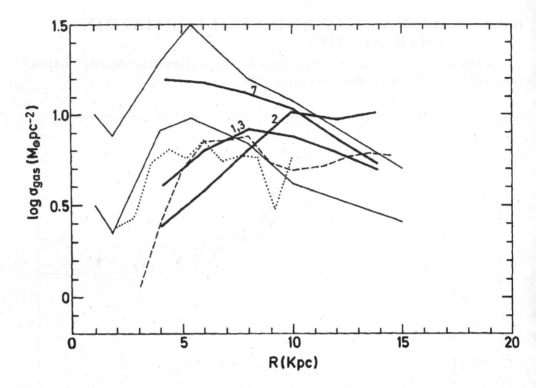

Figure 6.24. Radial distribution of the surface gas density (HI + H_2) by models with different SFRs as computed by Matteucci and François (1989) and compared with observations. Solid step-like curves show the data collected from Lacey and Fall (1985). The dotted line represents the observations of Talbot (1981). Short dashed line represents data from Bhat et al. (1984) and Li et al. (1982). The thicker lines represent the models. In particular, the curve labeled 2 refers to a model assuming the Schmidt law of star formation whereas the curve labeled 7 derives from a model adopting an exponent of the SF law $k = 1.1$. From Matteucci and François 1989, M.N.R.A.S. Vol. 239, 885; reproduced here by kind permission of Blackwell Science LTD (copy right 1989).

SFR can not vary too strongly with the surface gas density; for an exponent of the surface gas density $k = 2$ (Schmidt law) the gas distribution at the present time turns out to be too flat and even decreasing towards the inner regions of the disk at variance with observations, owing to the strong gas consumption required by this strong dependence on the gas density. In order to fit the gas distribution and the abundance gradients at the same time one should assume an exponent $k \leq 1.5$ and a dependence on the total surface mass density which is strongly declining towards the external disk regions. In Figure 6.24 the predictions for the gas distribution by various models assuming different SFRs are shown.

In Figure 6.25 are shown the predictions by the two-infall model, which takes into account the inside-out formation indicated by eq. 6.19 and a threshold in the SFR. Here the predictions from the best model with an exponent of the surface gas density $k = 1.5$ are shown and they reproduce much better the gas distribution along the disk than the equivalent models of Matteucci and François (see Figure 6.24); this is due to the assumed threshold in the SFR which regulates the gas consumption and avoids consumption of too much gas in the internal regions. The same threshold produces a quite flat gas distribution at large galactocentric radii keeping the gas level close to the threshold value.

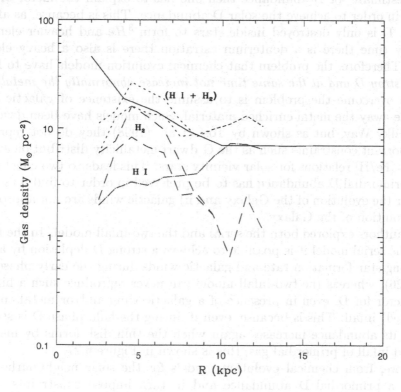

Figure 6.25. Radial gas distribution as predicted by the best model of Chiappini et al. (1997)(continuous line) compared to the observations (dashed, dotted and dashed-dotted lines). From Chiappini et al. 1997, Ap.J. Vol. 477, 765; reproduced here by kind permission of C. Chiappini and the University of Chicago Press (copy right 1997).

However, the gas distribution along the disk is not yet well understood and a final answer awaits chemo-dynamical models. Moreover, at present no model exists which is able to reproduce the molecular and the neutral hydrogen distributions, separately.

6.9 THE EVOLUTION OF LIGHT ELEMENTS AND THEIR IMPACT ON COSMOLOGY

In this section we will discuss the evolution of light elements (D, ^3He, ^4He, ^7Li, Be and B), namely those elements formed totally or partly during the Big Bang.

Let us start with deuterium which is entirely formed during the Big Bang and only destroyed in stars. As summarized in Table 1.5 there is no agreement on the primordial value of D, as measured by QSO absorption lines. If one believes the high estimate for D abundance then one has to explain the factor of ~ 10 depletion in order to achieve the solar D abundance. This is because, as already discussed, D is only destroyed inside stars to form ^3He and heavier elements, thus every time there is a deuterium astration there is also a heavy element increase. Therefore, the problem that chemical evolution models have to face is *how to destroy D and at the same time not increase abnormally the metallicity?* A way to overcome the problem is to assume the existence of galactic winds which take away the metal enriched material. Such models have been developed for the Milky Way, but as shown by Tosi et al. (1998) they do not reproduce other important constraints such as the G-dwarf metallicity distribution and the $[\alpha/Fe]$ vs. $[Fe/H]$ relations for solar vicinity stars. This leads to two conclusions: i) a low primordial D abundance has to be preferred in order to find reasonable models for the evolution of the Galaxy and ii) galactic winds are not appropriate for the evolution of the Galaxy.

These authors explored both the serial and the two-infall model. In the framework of the serial model it is possible to achieve a strong D depletion by assuming a strong star formation rate and galactic winds during the early phases (see Figure 6.26), whereas the two-infall model can never reproduce such a high depletion factor for D, even in presence of a galactic wind and/or metal enriched (D depleted) infall. This is because, even if during the halo phase D is strongly depleted, its abundance increases again when the thin disk forms by means of the second infall of primordial gas; this is shown in Figure 6.27.

Therefore, from chemical evolution models for the solar neighbourhood we can infer a primordial D abundance and in turn impose constraints on the baryon/photon ratio. From the models shown in Figure 6.27 it was concluded that D destruction is limited to a factor of 3 or less. If one adopts this as an upper bound then :

$$X_{2P} \leq 3X_{2ISM} \tag{6.22}$$

Adopting the Linsky et al. (1993) value for $(D/H)_{ISM} = 1.6 \pm 0.1 \cdot 10^{-5}$ and assuming for hydrogen $X = 0.70$ at the present time and $X = 0.76$ primordially, one can derive a 2σ upper bound of:

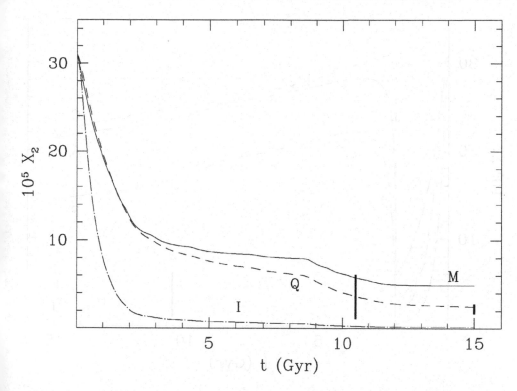

Figure 6.26. The evolution of D abundance in time in the framework of the serial model. The labels I, Q and M identify models with different IMFs. Model Q adopts a normal Scalo (1986) IMF. The assumed primordial abundance of D is $D_P = 3 \cdot 10^{-4}$ by mass. The D abundance at the epoch of the solar system formation and at the present time are reported. From Tosi et al. 1998, Ap.J. Vol. 498, 226; reproduced here by kind permission of M. Tosi and the University of Chicago Press (copy right 1998).

$$(D/H)_P \leq 5.0 \cdot 10^{-5} \qquad (6.23)$$

a value consistent with the low primordial D abundance derived by Tytler and collaborators. This upper bound to primordial D corresponds to a lower bound to the ratio η_{10} (baryon/photon ratio):

$$\eta_{10} \geq 4.0 \qquad (6.24)$$

or:

$$\Omega_B h^2 \geq 0.015 \qquad (6.25)$$

Such a lower bound to η_{10}, in the context of the standard Big Bang, leads to a predicted primordial mass fraction of ^4He ($Y_P \geq 0.244$) which is in modest disagreement with that inferred from observations of low-metallicity extragalactic HII regions, although systematic errors may be present in these data.

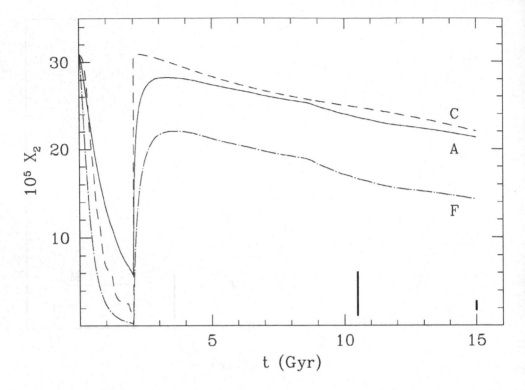

Figure 6.27. The evolution of D abundance in time in the framework of the two-infall model. The assumed primordial abundance of D is $D_P = 3 \cdot 10^{-4}$ by mass. The D abundance at the epoch of the solar system formation and at the present time are reported. Models A and C adopt a Scalo (1986) IMF whereas model F adopts a variable IMF. From Tosi et al. 1998, Ap.J. Vol. 498, 226; reproduced here by kind permission of M. Tosi and the University of Chicago Press (copy right 1998).

The abundance of D in the ISM is strictly connected to that of ^3He since D is transformed into ^3He inside stars. Unfortunately, most of models in the literature are unable to reproduce the abundance of ^3He at the time of the solar system formation and in the local ISM. All the models of Table 6.1, in fact, are overproducing ^3He and this is due to the excessively high ^3He yields adopted. Therefore, the problem should reside in the nucleosynthesis of ^3He in low and intermediate mass stars, the sites of production of this element. However, very recently new yields of ^3He from low mass stars have been computed and the predicted ^3He production has been substantially lowered (see chapter 2).

Concerning the evolution of ^4He, most models of chemical evolution assume a primordial He abundance of ~ 0.23 by mass, then the ^4He abundance increases during the galactic lifetime because it is produced by stars of all masses. Most of the ^4He should arise from intermediate to large mass stars, but also massive stars can substantially contribute to He- enrichment if a strong mass loss by stellar winds is taken into account (M92). Standard chemical evolution models with standard IMF (Salpeter or Scalo) and moderate mass loss by stellar winds predict an increase of He relative to metals over the galactic lifetime of at maximum $\Delta Y/\Delta Z = 2 - 3$, in agreement with independent estimates (Renzini 1994) based on the He content (Y=0.30-0.35) derived from the H-R diagram of stars in Baade's window. To increase this value up to the value suggested by observations of nearby stars ($\Delta Y/\Delta Z \sim 3$ Pagel and Portinari, 1998) or extragalactic HII regions ($\sim 4 - 6$ Pagel et al. 1992) one should adopt either some exotic IMF or assume that the metal production is inhibited by, for example, the formation of black holes instead of SN II in stars with masses larger than a certain mass (30-40 $M_\odot$, see Figure 6.28). In this figure the $\Delta Y/\Delta Z$ ratio is calculated as a function of the limiting mass for black hole formation. Two curves are shown, one corresponding to an early galactic age when the metallicity is $Z = 10^{-3}$ and the other refers to an age of 10.4 Gyr, when the solar metallicity is attained. The low metallicity curve indicates that a He-to-Z ratio inside the observational limits is obtained if stars more massive than $30 M_\odot$ are becoming black holes instead of SN II. The solar metallicity curve instead shows that this is obtained for much lower (probably unrealistic) values for this limiting mass ($15 - 17 M_\odot$). Clearly this low limiting mass would affect the evolution of other elements, and so one should be cautious in suggesting strong conclusions on this ground.

Lithium is produced during the Big Bang and there are indications that is also produced in stars. The observed log A(Li) vs. [Fe/H] (see Figures 1.20 and 6.29) shows that the Li abundance in Pop I and T-Tauri stars is at least a factor of ten higher than in Pop II stars.

The most common interpretation of this behaviour (see chapter 1) is that ^7Li has been produced during the galactic lifetime and that the uniform value of Pop II stars (the so-called "Spite-plateau") corresponds to the primordial ^7Li abundance. However, the existence of such a plateau has been challenged by some authors (e.g. Thornburn, 1994; Norris et al., 1994) who observed a trend of A(Li) vs. [Fe/H] in Pop II stars and suggested the following explanation: either a small production of ^7Li has occurred during the halo phase (cosmic ray α-α reactions) or some of the most metal poor stars have depleted their surface Li abundance; this second hypothesis is difficult to explain with the current models. A substantial stellar ^7Li production in the early Galaxy seems to be excluded by the detection of ^6Li in HD 84937 (Smith et al. 1993). A stellar production would, in fact, imply a very high $^7Li/^6Li$ ratio.

In any case, ^7Li should be produced in stars and one can envisage different categories of stars producing Li; among the suggested candidates we note the AGB stars, C-stars, type II SNe and novae. In particular, the AGB stars and C-stars are observed to be Li-rich (see chapter 1).

222

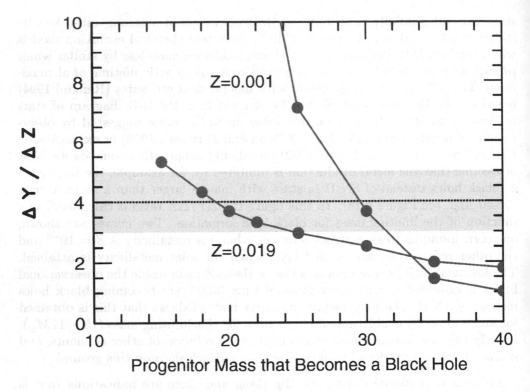

Figure 6.28. Helium-to-metal enrichment ratio as a function of the black hole mass limit. The observational limits are shown as the grey band. One curve refers to a time very early in the evolution when the total metallicity is $Z = 10^{-3}$ and another curve refers to a time t=10.4 Gyr when $Z = Z_\odot$. From Timmes et al. 1995, Ap.J. Suppl. 98, 617; reproduced here by kind permission of S.E. Woosley and the University of Chicago Press (copy right 1995).

An interesting feature of the A(Li) vs. [Fe/H] plot is that Pop II stars show a much smaller dispersion than Pop I stars. The smaller spread in Pop II stars can probably be explained as an effect of metallicity. D'Antona and Mazzitelli (1984), on the basis of stellar evolution calculations, suggested that main-sequence Li depletion is inhibited in low metallicity stars.

On the other hand, main-sequence Li destruction occurs in high metallicity stars because of deeper surface convection due to the higher opacity of Pop I stars. Therefore, the spread in Pop I stars is mainly due to age effects, the older stars being the most Li depleted.

In the *low* primordial Li abundance scenario one assumes that the Li abundance in Pop II stars reflects the ISM Li abundance at their birth, whereas only the Pop I stars of the upper envelope show the Li abundance at their birth. Therefore, models of galactic chemical evolution predicting the Li abundance in the ISM should be aimed at fitting the upper envelope of the A(Li) vs. [Fe/H] data.

In Figure 6.29 we show the predictions for the abundance of Li by assuming different stellar sources: AGB stars, C-stars, supernovae II and novae. The model including Li production from novae is the only one able to reproduce the steep increase of the Li abundance from the time of formation of the solar system up to now, as shown by the data.

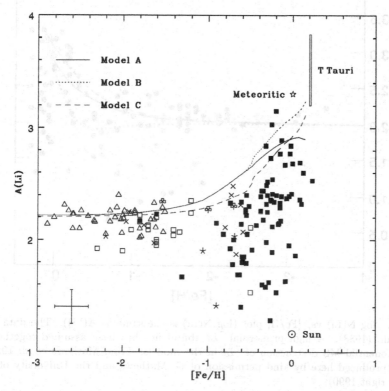

Figure 6.29. A(Li) vs. [Fe/H] plot. The data are the same as in Figure 1.20 with the addition of the abundances of TTauri stars, the meteoritic value and the Sun Li abundance. The models (continuous, dashed and dotted lines) represent the predictions of the two-infall model by assuming as a primordial Li abundance that of the plateau and different stellar sources of Li. The continuous line is the prediction of a model with AGB, C-stars and supernovae II as Li producers. The dotted line (the best model) includes novae as well as all the other sources. Finally, the dashed line represents a model which contains all the mentioned stellar sources but where the Li yields from SN II are assumed to be 1/2 of the yields assumed in the other two models. These models do not include the cosmic ray production of ^7Li but this should be not larger than $10-20\%$. Models from Romano et al. (1999). Courtesy of D. Romano.

Concerning ^7Li production from spallation the available data ($^7Li/^6Li$ ratio in the solar system and at the present time) do not allow yet any firm conclusion, as already discussed in chapter 1.

The alternative interpretation of a *high* primordial ^7Li abundance instead (that shown in Pop I stars) and ^7Li depletion in metal poor stars, would require an inhomogeneous Big Bang. This case has been explored by Mathews et al. (1990) and their results are shown in Figure 6.30.

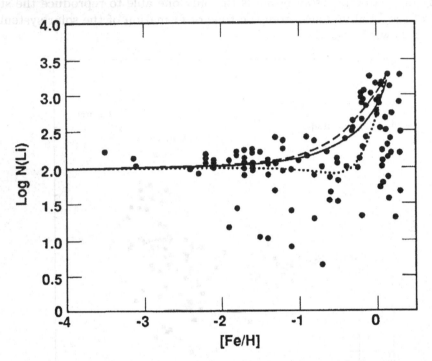

Figure 6.30. log N(Li) vs. [Fe/H] plot (log N(Li) is the same as A(Li)). The data are from Rebolo et al. (1988). A high primordial ^7Li abundance has been assumed together with a gradual exponential MS destruction of ^7Li in the models. From Mathews et al. 1990, Ap.J. 349, 449; reproduced here by kind permission of G. Mathews and the University of Chicago Press (copy right 1990).

This model shows that from the chemical evolution point of view it is not possible to distinguish among the *low* and the *high* primordial Li scenarios. However, the fact that Li-rich stars exist argues in favor of a stellar production of ^7Li.

^6Li, ^9Be and ^{11}B are believed to be produced by spallation reactions between cosmic rays and the ISM (see chapter 1). Spallation reactions may occur when p and α particles collide with the CNO atoms at rest in the ISM or when the fast CNO nuclei of cosmic rays collide with the p and α particles. Observations of ^9Be abundances show a linear correlation with [Fe/H] (see Figure 6.31). Ryan et al. (1990) and Gilmore et al. (1991) interpreted this linearity to be due to the fact that Be is synthesized in the immediate surroundings of SNe by spallation of fast nuclei of C and O ejected by the SNe against the protons of the ISM.

According to the time-delay model, this linearity may indicate that Be and Fe are produced on the same timescales.

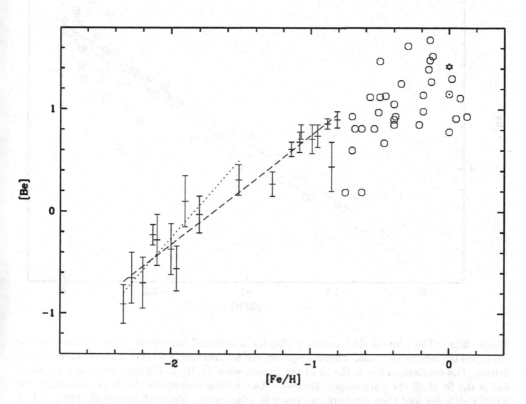

Figure 6.31. [Be] versus [Fe/H] for a selected sample of stars. The dashed line is the fit for stars with [Fe/H]<-0.9 dex, the dotted line is the fit for [Fe/H] <-1.4 dex. From Molaro et al. 1997, A & A Vol. 319, 593; reproduced here by kind permission of Springer Verlag (copy right 1997).

In Figure 6.32 is shown the abundance of [Be] plotted as a function of [O/H] by assuming that either [O/Fe] is constant in halo stars or that it increases with decreasing [Fe/H]. This increase of [O/Fe], if small, is more realistic than a completely flat plateau, since the [O/Fe] ratio produced by massive stars decreases with decreasing mass (see figures 6.6 and 6.7), and type Ia SNe, if originating from C-O white dwarfs may occur already at a galactic age of several 10^7 years. Therefore, a perfectly flat plateau of [O/Fe] in halo stars, as well as a very steep increase, should not be expected. As Figure 6.33 shows, the two different assumptions give rise to different correlations between [Be] and [O/H]. In particular, an increase of [O/Fe] with decreasing [Fe/H] would produce a steeper correlation.

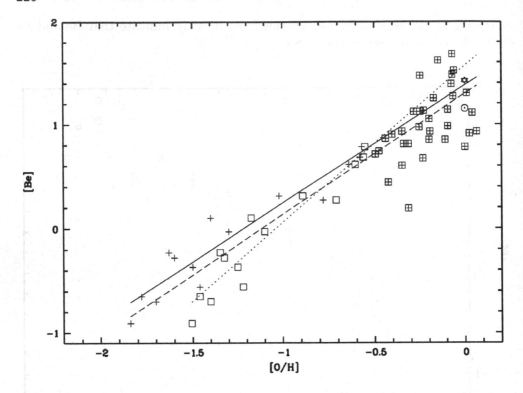

Figure 6.32. The plot of [Be] versus [O/H] for a selected sample of stars; crosses are for [O/Fe]=+0.5 dex in the halo, whereas squares are for increasing [O/Fe] towards lower metallicities. The continuous line is the fit of the crosses with [O/H] < -0.6 dex, whereas the dashed line is the fit of all the data sample (crosses). Dotted line represents the fit of the squares for [O/H]< -0.6 dex and then extrapolated towards solar values. From Molaro et al. 1997, A & A Vol. 319, 593; reproduced here by kind permission of Springer Verlag (copy right 1997).

6.10   THE CHEMICAL EVOLUTION OF THE GALACTIC BULGE

The evolution of the galactic bulge has not been as extensively studied as the disk or the halo although, as reviewed by Wyse and Gilmore (1992), different qualitative scenarios for the bulge formation have been proposed:

- the bulge formed by accretion of extant stellar systems which by dynamical friction eventually settled in the center of the Galaxy. Negligible star formation occurred *in situ*.

- The bulge formed from gas that had accumulated at the center of the Galaxy (either from gas-rich mergers or simply as a result of initial conditions) and evolved independently of the other galactic components with either i) a rapid star formation or ii) a slow star formation. A fast formation of the bulge with star formation in situ is favored by an interesting theoretical argument suggested by Elmegreen (1999). The argument goes as follows: the bulge potential well is too deep to have allowed a self-regulation mechanism or galactic winds triggered by the pressures from young stars and SNe, as occurs in disks and dwarf galaxies. Such a failure to self-regulate should lead to rapid and intense star formation on a timescale of only a few dynamical times. The gas accretion phase to form the bulge can be longer than this, but once a critical density is reached (threshold density), which depends primarily on the total virial density from dark matter, the star formation can be extremely fast.

- The bulge formed from the gas supply determined by inflow of metal enriched gas from the thick-disk or the halo and this process can be either rapid or slow.

- The bulge formed from the gas supply determined by inflow of metal enriched gas from the disk, this in turn is determined by the star formation timescale in the disk.

From the theoretical point of view few models for the chemical evolution of the galactic bulge have been presented in the last years: the first attempt to model the chemical evolution of the bulge was made by adopting the simple closed- box model (Rich 1988) with the conclusion that it fits very well the metallicity distribution of K giants with an effective yield of 1.8 $Z_\odot$. Then, the first detailed chemical evolution model for the galactic bulge was that of Matteucci and Brocato (1990); in this model the bulge was assumed to form by a fast collapse of primordial material and to suffer a rapid in situ star formation with an IMF flatter than in the solar vicinity. This was required to fit the K-giant metallicity distribution found by Rich (1988). This model predicted [α/Fe] > 0 for most of the bulge stars. As already discussed, this is the consequence of the time-delay model as opposed to the pseudo-secondary model.

Hensler et al. (1993) presented chemo-dynamical models where the bulge does not evolve separately from the other galactic components and forms from pre-enriched gas from the disk. However, no detailed predictions about abundance ratios were presented.

Molla and Ferrini (1995) adopted an IMF similar to that in the solar neighbourhood and a timescale of bulge formation relatively long ($\tau_{Bulge} = 1$ Gyr). They predicted, as expected, an evolution for the [α/Fe] ratios quite similar to that in the solar vicinity with < [O/Fe] >$_{Bulge}$∼ −0.15 dex.

Matteucci et al. (1999) proposed again for the bulge a fast in situ star formation from primordial gas, the same forming the galactic halo and collapsing faster into the center. The assumed efficiency of star formation in this model is higher than that in the solar neighbourhood, $\nu_{bulge} = 20\nu_\odot$ ($\nu_\odot = 0.5 Gyr^{-1}$) and the IMF is the Salpeter one ($x = 1.35$). The adoption of an IMF flatter than the Scalo one, which is good for the solar vicinity, is required to fit the metallicity distribution of bulge stars as measured by Mc William and Rich (1994) (see Figure 6.33 and also Figure 1.15).

As one can see, the data of Figure 6.33 seem to indicate that the best model, besides a SFR much faster than in the solar neighbourhood, requires the Salpeter IMF. In Figure 6.34 are shown the abundance ratios in the bulge predicted by the best model just mentioned. These ratios are compared to those in the solar vicinity. It is evident that under those conditions the $[\alpha/Fe]$ ratios are overabundant relative to the Sun for most of the metallicity range in bulge stars. Moreover, owing to the flatter IMF adopted, these overabundances are higher than those predicted for the halo stars in the solar vicinity.

However, the data for the bulge stars are still too uncertain to draw firm conclusions. We will only tentatively conclude that the available observational evidence and the most recent ideas on the star formation process indicate that the bulge formed very quickly and probably even faster than the inner halo. This point of view is obviously supported by the fact that most of the bulge stars are old, as discussed in chapter 1.

6.11 HOW RELIABLE ARE CHEMICAL EVOLUTION MODELS?

As we have seen in the previous chapters and sections, chemical evolution models make use of some basic approximations (such as I.R.A., S.M.L.A. and homogeneous and instantaneous mixing of gas) and adopt several more or less free parameters. The most important parameters are the stellar birthrate (IMF+SFR), the stellar nucleosynthesis and the possible gas flows.

We now discuss the reliability of the main approximations and parameters and whether we can envisage a unique model for the chemical evolution of the Milky Way.

6.11.1 BASIC APPROXIMATIONS

Let us start with the basic approximations:

- *The instantaneous recycling approximation (I.R.A.)*, i.e. to neglect the stellar lifetimes. This assumption is nowadays not used very much since it has became clear that the stellar lifetimes are necessary to follow the evolution of those chemical elements mainly produced by long-lived stars, such as Fe and N. Moreover, as shown by Prantzos and Aubert (1995), I.R.A. does not apply in situations where low gas fractions are obtained (e.g. in the inner regions of the galactic disk). In such cases, the large amounts of gas ejected by low mass stars at late times can modify substantially the final gas and abundances, also for elements produced on short timescales by massive stars. The differences between models with I.R.A. and without I.R.A. are shown in

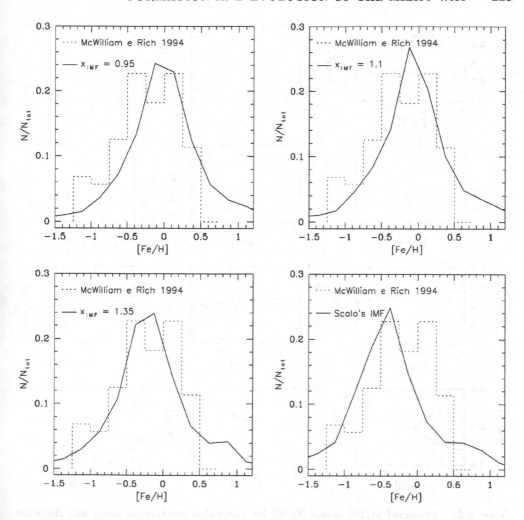

Figure 6.33. Predicted and observed metallicity distribution of bulge stars. The predictions are from Matteucci et al. (1999) and the data points from Mc William and Rich (1994). Each panel refers to a different IMF. From Matteucci et al. 1999, A & A Vol. 341, 458: reproduced here by kind permission of Springer Verlag (copy right 1999).

the figures 6.35 and 6.36, where the evolution of the gas fraction and metallicity as functions of time, and the metallicity as a function of the gas fraction are indicated, respectively. From these figures it is evident that models with I.R.A. strongly underestimate the amount of gas and overestimate the metallicity at late times. Concerning the evolution of the metallicity as a function of the gas fraction we can see that it does not increase monotonically in the non I.R.A. case, due to the dilution reached at late times.

230

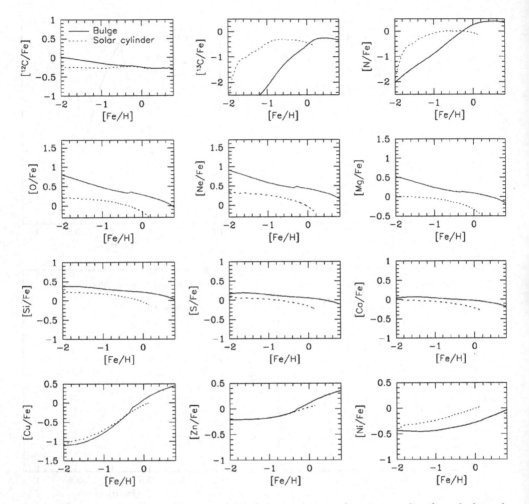

Figure 6.34. Predicted [α/Fe] versus [Fe/H] for the bulge (continuous lines) and the solar neighbourhood (dotted lines). From Matteucci et al. 1999, A & A Vol. 341, 458: reproduced here by kind permission of Springer Verlag (copy right 1999).

From the discussion above we conclude that the I.R.A. approximation can still be valid if one wants to follow the evolution of the global metallicity Z dominated by oxygen which is produced on very short timescales and avoids modelling regions where the gas fraction is lower than 0.1.

- *The instantaneous and homogeneous mixing of gas.* This approximation is still widely adopted in chemical evolution models; it means that the ejecta from the stars are assumed to mix instantaneously and homogeneously with the gas present in the region one is modelling. This assumption sounds particularly weak if one applies it to a large spatial region, whereas it could be correct

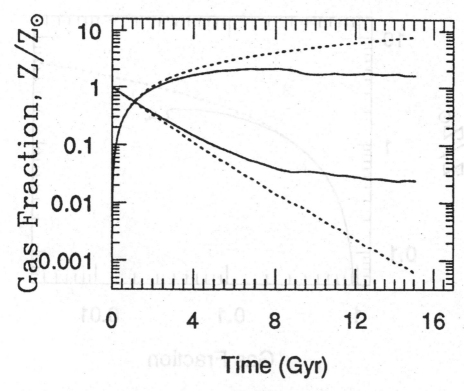

Figure 6.35. Evolution of the gas fraction (lower curves) and metallicity relative to the solar one (upper curves) in models with I.R.A. (dotted lines) and without I.R.A. (solid lines). I.R.A. is a good approximation as long as the gas fraction is > 0.1. From Prantzos and Aubert 1995, A & A Vol. 302, 69; reproduced here by kind permission of Springer Verlag (copy right 1995).

for a relatively small region. For example, if one is studying the evolution of the solar vicinity, then the assumption requires that the stellar ejecta mix with all the gas present in the whole solar neighbourhood at any time. This is probably a too simplistic approximation, and the reason for that is the observed spread in abundances and abundance ratios especially in halo stars, at low metallicities.

Some authors have attempted to adopt models relaxing this assumption or, at least modelling the chemical evolution of smaller regions and assuming no mixing between them (e.g. Malinie et al. 1993; Copi 1997). In the Copi model, 1000 different galactic regions of $10^5 M_\odot$ (remember that the total mass in the solar neighbourhood is of the order of $10^8 M_\odot$) are evolved at the same time but the history of the star formation is slightly different in each region; this is obtained with a Monte Carlo technique. The model applied to each region is either closed or with infall or outflow. In the case

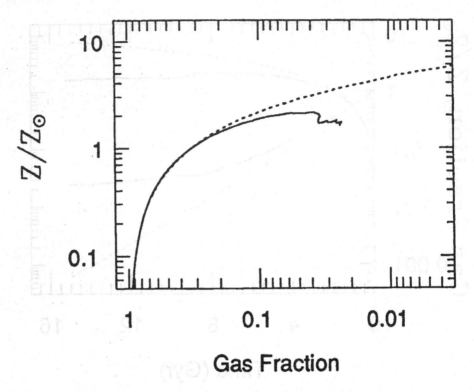

$Z/Z_\odot$

Gas Fraction

Figure 6.36. Evolution of the metallicity relative to the solar one as a function of the gas fraction from Figure 6.35. *Dotted line*: model with I.R.A.; *solid line*: model without I.R.A. There is a factor of ~ 2 discrepancy between the two curves for a gas fraction ~ 0.02, becoming larger for even lower values of the gas fraction. From Prantzos and Aubert 1995, A & A Vol. 302, 69; reproduced here by kind permission of Springer Verlag (copy right 1995).

of the outflow model, only the products of SN Ia and II are ejected into the intergalactic medium. The distribution of the abundances obtained in this way shows an obvious spread. For example in Figure 6.37 we show the predicted age-metallicity relations compared with the data of Edvardsson et al. (1993), while in Figure 6.38 we report the predictions for the [O/Fe] versus [Fe/H] relations. The spread in the figures is also due to the fact that the predictions of different models (closed, infall, outflow) are plotted together with the predictions of the same model but in different spatial regions.

The chemo-dynamical models adopting the SPH method (see chapter 5), also predict the existence of a spread in the relative abundances; this is due to the fact that gas, stars and dark matter are treated as an ensemble of discrete particles. As an example of this kind of models we show the prediction of the [O/Fe] versus [Fe/H] ratio obtained by the abundances predicted for the different star particles (see Figure 6.39). In the plot are shown all the stars located at galactocentric distances $R_G > 2$ kpc which indicate the existence of a trend similar to the observed one but with a large intrinsic spread. However, these results should be still taken with care since they predict that at the center of the Galaxy the majority of stars are young and with low [O/Fe], at variance with observations (see sections 1.7 and 6.10).

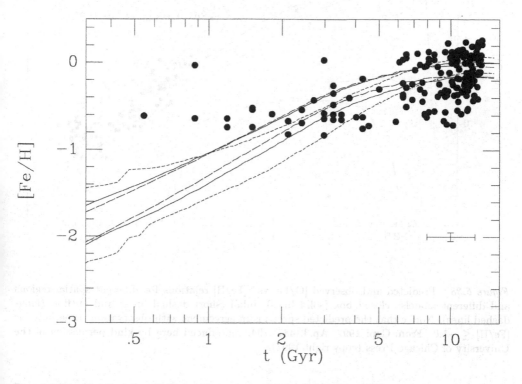

Figure 6.37. Predicted and observed age-metallicity relations for different spatial regions and different models: closed box model (solid lines), infall model (short-dashed lines), outflow model (long-dashed lines). The data are from Edvardsson et al. (1993) and a typical error bar is shown in the bottom right corner of the figure. From Copi 1997, Ap.J. 487, 704; reproduced here by kind permission of the University of Chicago Press (copy right 1997).

From the physical point of view, it is not yet clear how mixing should affect galaxy evolution and, from what we know, we would expect more mixing than is found. Addressing this point we report the reasoning given by Edmunds (1975) on how to estimate the expected abundance spread in the presence of efficient mixing in the ISM. The reasoning is as follows: let us start considering fluctuations by a factor of two corresponding to $\delta(O/H) \sim 3 \cdot 10^{-4}$ (abundance by number), which is the difference between the Sun and Orion nebula or the difference observed between HII region abundances over a scale length of 1 kpc along the disk, and see if these fluctuations are compatible with a perfectly mixed

234

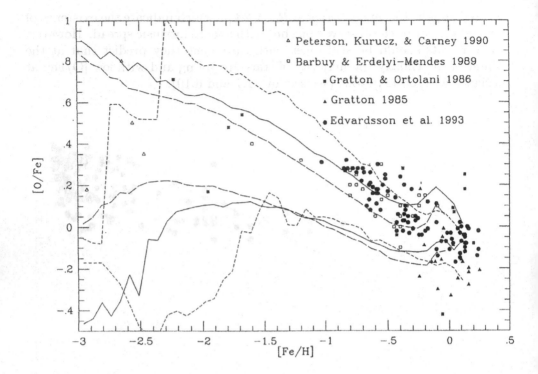

Figure 6.38. Predicted and observed [O/Fe] vs. [Fe/H] relations for different spatial regions and different models: closed box (solid lines), infall (short-dashed lines) and outflow (long-dashed lines). Notice that the predicted spread is in agreement with observations especially for [Fe/H] ≤ -1.0. From Copi 1997, Ap.J. 487, 704; reproduced here by kind permission of the University of Chicago Press (copy right 1997).

disk. Let us assume that massive stars are randomly distributed in space and time in order to derive what would be the expected abundance fluctuations in a fully mixed ISM enriched only by isolated stars. We assume that every massive star develops a wind-blown bubble into which it finally explodes as a supernova. The winds and explosions act as an effective mixing process. If each SN remnant reaches a radius of about 50 pc before it becomes indistinguishable from the ISM and the disk has a maximum radius $R_{max}=12$ kpc and height of 200 pc, then each bubble and SN remnant will occupy a fractional volume of $6 \cdot 10^{-6}$ of the disk. Assuming for simplicity a constant SN rate of $1/100$ yr over 10^{10} years, then at any point of the disk an average of 580 SN events would have contributed by now to the metal enrichment of the ISM. We suppose that the metals are well mixed within the 50 pc radius during the lifetime of a remnant and that the production of SNe is randomly distributed through the disk. If n events contribute to give approximately the oxygen solar abundance by number

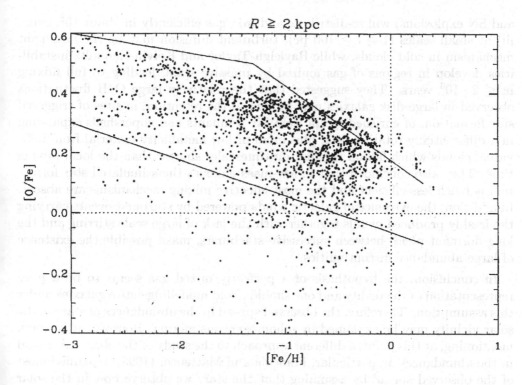

Figure 6.39. Predicted [O/Fe] vs. [Fe/H] for the star particles in the model of Raiteri et al. (1996). The star particles refer only to stars located at galactocentric distances > 2 kpc. The solid lines indicate the range in which the observational data are located. From Raiteri et al. 1996, A. & A Vol. 315, 105; reproduced here by kind permission of Springer Verlag (copy right 1996).

of $6.9 \cdot 10^{-4}$, then fluctuations of $\delta(O/H) \sim (O/H)_\odot / n^{1/2} \sim 3 \cdot 10^{-5}$ are obtained, a value which is at least ten times lower than observed. This discrepancy between the predicted fluctuations and the observed ones can be diminished by relaxing some of the extremely simplifying assumptions made here, such as the random massive star formation.

Roy and Kunth (1995) discussed the problem of mixing in the ISM in great detail and suggested three main mixing mechanisms acting on different spatial scales and, on the basis of that, suggested that again the ISM should appear more chemically homogeneous than it is. In particular, their conclusions were: i) on large galactic scales ($1 > l \geq 10$ kpc), turbulent diffusion of interstellar clouds in the shear flow of galactic differential rotation is able to wipe out azimuthal O/H fluctuations in less than 10^9 years; ii) at the intermediate scale ($100 \geq l \geq 1000$ pc), cloud collisions and expanding supershells driven by evolving massive star associations, differential rotation and triggered star formation (by stellar winds and SN explosions) will re-distribute and mix gas efficiently in about 10^8 years; iii) at small scales ($1 \geq l \geq 100$ pc), turbulent diffusion may be the dominant mechanism in cold clouds, while Rayleigh-Taylor and Kelvin-Helmoltz instabilities develop in regions of gas ionized by massive stars, leading to full mixing in $\leq 2 \cdot 10^6$ years. They suggested that the relatively large O/H fluctuations observed in large disk galaxies may be due to the retention, in sites of triggered star formation, of enriched ejecta from SN remnants and supershells expanding in a differentially rotating disk, plus infall of low metallicity material from individual clouds which fall on the disk on timescales shorter than the local mixing time. They also concluded that in low-mass galaxies the stimulated star formation is much less efficient and the most effective mixing mechanisms are absent. In addition, the existence of galactic winds powered by starburst events carrying the freshly produced metals together with the lack of large scale stirring and the long dormant phase between successive starbursts, make possible the existence of large abundance discontinuities.

In conclusion, the hypothesis of a perfectly mixed gas seems to be a poor representation of the reality and one should avoid modelling entire galaxies under this assumption. Therefore, the observed spread in the abundances of stars in the solar vicinity may be explained by inhomogeneous mixing. However, it is worth mentioning, at this point, a different approach to the study of the observed spread in the abundances; in particular, François and Matteucci (1993) explained most of the observed spread by assuming that the stars we observe now in the solar neighbourhood were not all born there but some were scattered into the solar neighbourhood from other galactic regions because of orbital diffusion. Since different regions of the disk are likely to evolve at different rates (different star formation and infall rates) one expects abundance gradients, which are indeed observed in the youngest objects, such as HII regions and B stars, which did not have the time to be diffused into other regions. The abundance gradient coupled with the orbital diffusion then produces the observed scatter. Since the orbital diffusion is stronger for older stars, one expects that as [Fe/H] decreases there is an increase in the range of radii at which a star, now seen close to the Sun, can have formed. François and Matteucci modeled this increasing range of galactocentric distances possible for a star of a given mass and metallicity on the basis of the orbital diffusion theory. They found a R_{max} and a R_{min} for each star visible now in the solar vicinity and then by means of a chemical evolution model they computed the $[\alpha/Fe]_{min}$, $[\alpha/Fe]_{max}$, $[Fe/H]_{max}$ and the $[Fe/H]_{min}$. The spread was then given by the differences between the max and min values.

In Figure 6.40 we show the spread predicted for [Mg/Fe] vs [Fe/H] by this model whereas in Figure 6.41 is shown the predicted spread for the age-metallicity relation. In each figure are shown also the predictions for the galactic bulge since some bulge stars could also be seen now in the solar neighbourhood (Grenon 1987).

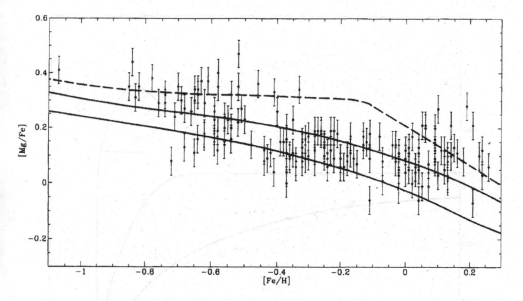

Figure 6.40. The plot of observed [Mg/Fe] vs. [Fe/H] from Edvardsson et al. (1993) is compared with the models by François and Matteucci (1993) (continuous lines), as described in the text. The predictions for the galactic bulge from Matteucci and Brocato (1990) are also shown (dashed line). From François and Matteucci 1993, A & A Vol. 280, 136; reproduced here by kind permission of Springer Verlag (copy right 1993).

6.11.2   THE UNIQUENESS

Before concluding this section on the reliability of chemical evolution models we mention a study made by Tosi (1988) who performed a detailed analysis of the influence of the various parameters (SFR, IMF, infall etc..) on the chemical evolution of galaxies and discussed the problem of the uniqueness of models. She concluded that only a few (but more than one) models reproduce the majority of the observational constraints in the Milky Way and that this is possible under the following assumptions:

- a) the SFR has not decreased rapidly with time;

- b) the SFR is not simply proportional to the gas density;

- c) the IMF has not varied strongly in space and time and does not imply too large a fraction of massive stars;

- d) the rate of gas infall should decrease in time more slowly than the SFR and its present value should be in the range $0.3 - 1.8 M_\odot yr^{-1}$;

238

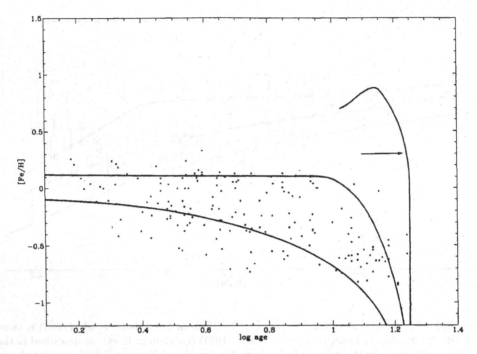

Figure 6.41. The age-metallicity relation by Edvardsson et al. (1993) compared with the model predictions of François and Matteucci (1993) (continuous lines), as described in the text. The predictions for the galactic bulge from Matteucci and Brocato (1990) are also shown (the line indicated by the arrow). From François and Matteucci 1993, A & A Vol. 280, 136; reproduced here by kind permission of Springer Verlag (copy right 1993).

- e) under the very simple assumption that radial flows inward in the disk carry a fraction of the infalling gas, the conclusion is that in order to reproduce the majority of observational constraints this fraction should not exceed 50%.

In spite of the fact that these conclusions were based on the observational data available in 1988, they are still valid in their general meaning.

She also showed that model predictions of elemental and isotopic ratios depend mostly on the adopted IMF and stellar nucleosynthesis, model predictions on radial distributions and age-metallicity relations depend mostly on the adopted SFR, IR and IMF, and model predictions of the absolute abundances depend on the combination of all these parameters.

Later on, Prantzos and Aubert (1995) made a similar experiment by adopting several laws for the SFR and the infall rate and comparing the results with a *minimal set* of observational constraints available at that time. They did not consider radial flows in their models. As a general rule, one should remember that a good model of chemical evolution should reproduce a number of observables larger than the number of free parameters in the model!

Their conclusions can be summarized as follows:

- i) a simple radial dependence of the SFR such as $SFR(R_G) \propto R_G^{-1} \sigma_{gas}^1$ can account for most of the observational features of the galactic disk (abundance gradients, gas and SFR distributions);

- ii) they showed in a quantitative way (see Figures 6.35 and 6.36) the danger of adopting I.R.A. in chemical evolution models, especially if applied to the inner galactic regions;

- iii) an exponentially declining infall rate $f(t) \propto e^{(-t/\tau)}$, with $\tau = 3$ Gyr constant over the disk was suggested.

In this case, two of the conclusions, i) and iii), are no more valid. In fact, they were both based on a very shallow abundance gradient along the disk claimed by the first studies of B stars (Kaufer et al. 1994). More recent studies of B stars, as we discussed before, suggest instead a gradient for oxygen as steep as that measured from HII regions. In such a case, a timescale for the formation of the disk increasing with galactocentric distance and a stronger dependence of the SFR on the gas density (namely $k = 1.5$) are required to match the observations, as we have already discussed.

In conclusion, we have not yet identified *the unique model* for the chemical evolution of the Milky Way, but owing to the work of these last 20 years we have understood how the different parameters influence the chemical evolution, and that the most crucial one is the stellar birthrate (IMF+SFR) and in particular the initial mass function. To this purpose it is worth recalling the First Commandment of galactic chemical evolution, as expressed by M. Edmunds: *Thou shalt not alter the initial mass function for star formation- except in an emergency.*

We now appreciate that abundance ratios represent the most reliable predictions that we can make, and that they are powerful tools to date and identify high-redshift objects.

On the basis of our knowledge at the time of writing we can say that models of chemical evolution of the Milky Way are probably reliable enough to suggest the following picture for its formation and evolution:

the inner halo and the bulge probably formed out of the same gas on similar timescales of the order of 0.5-2 Gyr. The outer halo formed perhaps on a longer timescale (3-5 Gyr) by accretion of extragalactic gas or by accretion of stellar systems such as dwarf spheroidals and dwarf irregulars. The thick and the thin-disk very probably formed out of gas different from that forming the halo and bulge. The thin-disk, in particular, formed by means of an inside-out mechanism with the innermost regions forming quite fast and the outermost regions still forming now. The disk in the solar neighbourhood must have assembled on a relatively long timescale (larger than 6 Gyr). The SFR in the halo and thick-disk must have been faster than in the thin-disk and there are some indications from the [Fe/O] vs. [O/H] and [Fe/Mg] vs. [Mg/H] relations that the star formation could have stopped for a certain period of time in between the formation of the thick- and the thin-disk. The SFR in the disk did not change much in time although we do not yet understand the precise star formation law, but it changed in space in the sense that it was stronger in the innermost than in the outermost regions of the disk. The best star formation law seems to be depending on both the surface gas density and the total surface mass density or to have an efficiency inversely proportional to the galactocentric distance. There is no evidence for a strong variation of the IMF in space and time neither observationally nor theoretically, although one should expect that the conditions for star formation would be sensitive to the environmental conditions.

Finally, the contribution to the chemical enrichment in Fe from SN Ia in our Galaxy must have been substantial (70 − 80%) in order to explain most of the current data on [α/Fe] vs. [Fe/H] on the basis of the time-delay model, the IMF derived for the solar neighbourhood and the available nucleosynthesis calculations.

A lot of effort is still needed, especially in understanding the physics of the star formation and the link between gas dynamics and chemical evolution.

Chapter 7

NUCLEOCHRONOLOGY AND THE AGE OF THE GALAXY

7.1 RADIOACTIVE DATING OF ELEMENTS

Radioactive dating of elements consists in using the meteoritic abundances of the decay products of certain radioactive elements to derive information on their age distribution in the solar system. This information, in turn, is relevant to the duration of the star formation and nucleosynthesis history before the Sun formed and therefore to the age of the Galaxy. In particular, nucleochronology utilizes our knowledge of the abundances and production rates of radioactive nuclear species in stars in the framework of chemical evolution models.

7.1.1 ABUNDANCES IN METEORITES

Generally, the interesting chemical elements in connection with the determination of the galactic age are long-lived radioactive isotopes, which are shown in Table 7.1. In this Table, taken from Cowan et al. (1991), are shown the potentially interesting nuclear chronometers (with lifetimes $> 10^7$ years), their decay times, their daughters and the nucleosynthesis processes from which they are thought to originate.

These chronometers, with the exception of ^{40}K, are all heavy nuclei ($A > 60$) and are products of r-process, s-process or p-process nucleosynthesis. As briefly mentioned in chapter 2, s- and r-process refers to the slow or rapid neutron capture by seed nuclei relative to the timescale of the β-decay. Concerning the p-process elements they are proton-rich elements (from ^{74}Se to ^{196}Hg) which can be explained neither with s- nor with r-processes. They are believed to originate from reactions of the type (p, γ) and (γ, n) on s-process elements. The mechanism of s-process neutron capture requires an appropriate source of free neutrons and some seed nuclei by which they are captured. The seed nuclei consist of iron-peak elements, while neutrons are provided by reactions such as $^{13}C(\alpha, n)^{16}O$ and $^{22}Ne(\alpha, n)^{25}Mg$ which are active during the evolution of massive stars and red giant stars. In particular, the s-process isotopes up to mass A=90 are synthesized

241

Table 7.1.   Long-lived galactic chronometers

Nucleus	Decay time (years)	Daughter	Nucleosynthesis Process
^{40}K	$1.3 \cdot 10^9$	^{40}Ca	s-/ ?
^{40}K	$1.3 \cdot 10^9$	^{40}Ar	s-/?
^{87}Rb	$4.9 \cdot 10^{10}$	^{87}Sr	r-/s-process
^{92}Nb	$3.7 \cdot 10^7$	^{92}Zr	p-process
^{107}Pd	$6.5 \cdot 10^6$	^{107}Ag	r-/s-process
^{129}I	$1.6 \cdot 10^7$	^{129}Xe	r-process
^{138}La	$1.1 \cdot 10^{11}$	^{138}Ba	p-process
^{138}La	$1.1 \cdot 10^{11}$	^{138}Ce	p-process
^{146}Sm	10^8	^{142}Nd	p-process
^{147}Sm	$1.1 \cdot 10^{11}$	^{143}Nd	s-/r-process
^{176}Lu	$3.7 \cdot 10^{10}$	^{176}Hf	s-process
^{187}Re	$4.5 \cdot 10^{10}$	^{187}Os	r-process
^{205}Pb	$1.5 \cdot 10^7$	^{205}Tl	s-process
^{232}Th	$1.4 \cdot 10^{10}$	^{208}Pb	r-process
^{235}U	$7.0 \cdot 10^8$	^{207}Pb	r-process
^{236}U	$2.3 \cdot 10^7$	^{232}Th	r-process
^{238}U	$4.5 \cdot 10^9$	^{206}Pb	r-process
^{244}Pu	$8.2 \cdot 10^7$	^{232}Th	r-process
^{247}Cm	$1.6 \cdot 10^7$	^{235}U	r-process

in massive stars ($M > 10 M_\odot$) during the He-burning phase, whereas the heavier s-process elements are produced during the He-burning in the shell in the thermal pulse phase. On the other hand, the r-process elements are produced during explosive events such as supernovae.

The most important radioactive chronometers are r-process elements, in particular the pairs $^{232}Th/^{238}U$, $^{235}U/^{238}U$ and $^{187}Re/^{187}Os$. Therefore, it is important to know the history of r-process synthesis over the lifetime of the Galaxy in order to compute the evolution of the abundances of such elements.

7.1.2   BASIC PRINCIPLES OF NUCLEOCHRONOLOGY

Since many important reviews on this subject have appeared in the past years (e.g. see Tinsley 1980 and references therein and Cowan et al. 1991) we recall here only few basic principles of nucleochronology. The main idea is simple: one computes the temporal evolution of the abundances of the relevant nucleids in the gas and then compares these abundances and their ratios with those derived for the solar system (from the abundances of decay products in meteorites). In this way, one can obtain the age corresponding to the formation of the solar system, t_{SS}, as the time for which the predicted abundances reproduce the observed ones.

The simplest situation that one can imagine is to assume that all the elements formed in an instantaneous burst of star formation, then their decay would lead to abundances at a time t in the ISM given by:

$$X_i(t) = X_{io}e^{-\lambda_i t} \tag{7.1}$$

where X_{io} is the initial abundance of the radioactive element i and is proportional to its yield y_i. In particular, at the time t_{SS} (the time of the formation of the solar system), two radioactive elements would have the following abundance ratio:

$$\frac{X_i(t_{SS})}{X_j(t_{SS})} = \frac{X_{io}}{X_{jo}}e^{-(\lambda_i - \lambda_j)t_{SS}}, \tag{7.2}$$

where $X_{io}/X_{jo} = y_i/y_j$ and λ_i and λ_j are the decay constants. In this way, the value of t_{SS} can be found by knowing the yields of the two elements (in other words their production ratio) and the measured abundance ratio in the solar system. Then, knowing that the age of the solar system is ~ 4.5 Gyr one can obtain the lifetime of the Galaxy by:

$$t_G = t_{SS} + 4.5 \quad Gyr \tag{7.3}$$

This estimate is of course quite unrealistic since the chemical elements have been continuously manufactured during the period of time elapsed from the first star formation to the formation of the solar system, t_{SS}, and therefore one needs to consider a more detailed approach.

One can start from analytical models of galactic chemical evolution and recall the basic equation for a given chemical element i for the case with both infall and outflow (eq. 5.30). This equation should be modified to account for the radioactive decay:

$$\frac{d(X_i M_{gas})}{dt} = -X_i\psi(t) + E_i(t) - \lambda_i X_i M_{gas} + X_{iA}A(t) - X_i W(t) \tag{7.4}$$

where:

$$E_i(t) = \int_{m(t)}^{\infty} [(m - M_R)X_i(t - \tau_m)e^{-\lambda_i \tau_m} + mp_{im}]\psi(t - \tau_m)\varphi(m)dm \tag{7.5}$$

The decay timescale λ_i^{-1} is obviously independent of the timescales for star formation and infall/outflow and there are straightforward analytical solutions for X_i only in special cases, even if the I.R.A. is adopted (Tinsley, 1980). On the other hand, I.R.A. is meaningless in computing the evolution of radioactive elements if they are partly or entirely produced by long-lived stars.

Clayton (1985;1988) has shown that under specific cases for the SFR and IR, one can find an analytical solution also for radioactive elements, in particular in analogy with eq. (5.55):

$$(X_i - X_{iA}) = (y_Z \omega - \lambda X_{iA}) e^{-\lambda_i t} (\frac{\Delta}{t + \Delta})^k I_k(t, \lambda_i) \tag{7.6}$$

where :

$$I_k(t, \lambda_i) = \int_0^t \frac{t' + \Delta}{\Delta} e^{-\lambda_i t'} dt' \tag{7.7}$$

is a function depending on k, Δ and ω, which have been already defined in chapter 5 when discussing Clayton's *Standard Model*. We note again that ω is related to the SFR, while Δ and k are related to the infall rate (see eq. 5.54).

If the infall abundance $X_{iA} << X_i$, as is reasonable for a radioactive element, one can set $X_{iA} = 0$ and divide eq. (7.6) by eq. (5.55) for X_i where $X_{io} = 0$, and obtain the ratio $r(t)$ between the concentration of the radioactive element i and the concentration it would have if it were stable, the so-called *remaining fraction*:

$$r(t) = \frac{X_i(t, \lambda_i)}{X_i(t, \lambda_i = 0)} = \frac{k+1}{\Delta} \frac{e^{-\lambda_i t} [\Delta/(t+\Delta)]^k I_k(t, \lambda_i)}{(t+\Delta)/\Delta - [(t+\Delta)/\Delta]^{-k}}. \tag{7.8}$$

The fascinating feature of eq. (7.8) is that it does not depend upon ω, which is related to the specific star formation history, but it depends only on the parameters k and Δ. In this way we can express the ratio between two radioactive isotopes as:

$$\frac{X_i}{X_j} = \frac{X_i/X_i(\lambda_i = 0)}{X_j/X_j(\lambda_j = 0)} \frac{X_i(\lambda_i = 0)}{X_j(\lambda_j = 0)} \tag{7.9}$$

and then, by introducing the remaining fractions and the yields, we obtain:

$$\frac{X_i(\lambda_i = 0)}{X_j(\lambda_j = 0)} = \frac{y_i}{y_j} \tag{7.10}$$

we have:

$$\frac{X_i}{X_j} = \frac{r_i(t)}{r_j(t)} \frac{y_i}{y_j} \tag{7.11}$$

Clearly the case $k = 0$ corresponds again to the simple model with no infall and the remaining fraction in this case is:

$$r_o = (1 - e^{-\lambda_i t})/\lambda_i t \tag{7.12}$$

Therefore, once the parameters k and Δ are chosen and this can be done on the basis of other constraints to the chemical evolution model, one needs to know the ratio between the meteoritic abundances of the elements of interest and their production ratio, which is equivalent to the ratio of their yields, in order to derive t_{SS}. Another advantage of this formulation is that it allows us to impose constraints on the infall law through the nucleochronology, as long as we trust nucleosynthesis calculations and meteoritic abundances.

However, all of these analytical solutions have been derived under I.R.A. and therefore the same criticism that we applied to models with I.R.A. can be applied now, especially if one of the chronometers is an s-process element produced by long-lived stars. In such a case a detailed numerical model should be adopted.

Concerning the meteoritic abundances, what one really measures is the present time abundance ratios of radioactive nuclei and therefore one needs to derive the ratio at the time of the formation of the solar system by using the relative lifetimes for the decay. For example, Anders and Grevesse (1989) using various meteorite samples found the present day value of Th/U=3.6 which translates into $^{232}Th/^{238}U$ =2.32 at the time of formation of the solar system. For the ratio $^{235}U/^{238}U$ at the formation of the solar system these authors give a value of 0.317. These abundances are typically accurate to about $5 - 10\%$.

7.2   THE AGE OF THE GALAXY
7.2.1   GALACTIC AGE FROM CHRONOMETERS

Many authors have adopted different chronometers (mainly r-process elements) to derive the age of the Galaxy based on the principles outlined above and we report some of the estimates of the galactic age in Table 7.2. Taking all of these estimates together we can envisage a range for t_G of 11-14 Gyr. However, if we disregard results obtained with the closed-box model, which does not represent a good approximation to the evolution of the local disk, we find age determinations in the range 13-15 Gyr, although values in the broader range 10-20 Gyr cannot be excluded owing to the uncertainties still existing in the chemical evolution models and nucleosynthesis production rates. Such uncertainties include the fact that some chronometers such as ^{143}Nd are partly s- and partly r-process elements and this complicates the calculation of the evolution of their abundances with time.

A more direct way of using the Th/Nd ratio is to measure the strength of the Th and Nd lines in a sample of nearby (mostly dwarf) stars (Butcher, 1987). He found no variation of the Th/Nd ratio with the stellar age, and since ^{232}Th has an half-life of $1.4 \cdot 10^{10}$ years, this implies that the age of the galactic disk is shorter than this time. In particular, he derived an age ≤ 9.6 Gyr for the disk (see Table 7.2) under the assumption that the growth rate of the two elements is the same. However, he did not take into account the fact that Th is an r-process element while Nd derives partly from r- and partly from s-process. This makes this estimate quite uncertain, since one has to take into account the

Table 7.2.   Chronometric age determinations

Chronometer	Age (Gyr)	Reference
Th/U	$8.4 \leq t_G \leq 11.6$	Fowler & Meisl (1986), Fowler (1987)
Th/U	$9 \leq t_G \leq 28$	Meyer & Schramm (1986)
Th/U	$12.4 \leq t_G \leq 14.7$	Cowan et al. (1987)
Th/U	$t_G < 16$	Clayton (1988)
Th/U	$13.2 \leq t_G \leq 15.2$	Thielemann et al. (1983)
Th/U	$t_G = 15.6 \pm 4.6$	Cowan et al. (1999)
Re/Os	$11 \leq t_G \leq 15$	Yokoi et al. (1983)
Re/Os	$14 \leq t_G \leq 20$	Clayton (1988)
Pb	$12 \leq t_G$	Beer (1990)
Th/Nd	$t_G \leq 9.6$	Butcher (1987)
Th/Nd	$t_G \leq 20.0$	Mathews & Schramm (1988)
Th/Nd	$t_G \leq 20.0$	Clayton (1988)
Th/Nd	$t_G \leq 13$	Malaney et al. (1989)
Th/Nd	$15 \leq t_G \leq 20$	Lawler et al. (1990)

correct chemical evolution of these elements before a reliable age of the Galaxy can be derived. Therefore, I will conclude this section with the words of Clayton (1988): *only a detailed and specific and correct model for the growth and chemical evolution of the solar neighbourhood can enable the galactic age to be inferred from radioactivity.*

7.2.2   OTHER AGE ESTIMATES

As discussed in chapter 6, the galactic globular clusters represent the oldest objects, and therefore their age, if known, can give us the age of the Galaxy. Previous age determinations for globular clusters suggested a range of t_G from 12 to 21 Gyr (see Cowan et al. 1991 for a review) but more recent determinations based on distances derived from data taken with the Hipparcos satellite seem to indicate the lower age limit, in particular $t_G = 11.8^{+2.1}_{-2.5}$ Gyr (see Table 7.3).

A completely independent way of deriving the age of the Galaxy uses the study of the luminosity function of the white dwarfs. However, the age derived in this way refers only to the galactic disk. This method was first suggested by Schmidt (1959) and its physical basis is explained by the fact that white dwarfs are cooling and their cooling is faster if they are hotter. The space density of white dwarfs is therefore expected to increase monotonically with decreasing white dwarf luminosity. In fact, a fall-off in the luminosity function of white dwarfs is indeed observed. The fall-off corresponds roughly to a luminosity of $\log(L/L_\odot) = -4.5$ and is presumably due to the finite age of our Galaxy. The determination of this age then depends on models for the cooling histories of the white dwarfs. For a review on the subject the reader is addressed to the paper of D'Antona and Mazzitelli (1990) where a detailed discussion of theoretical luminosity functions of white dwarfs can be found. For our purposes we recall only that the various studies suggest an age of the galactic disk of 9-10 Gyr. It is worth noticing that this age is probably not the age of the Galaxy, in the sense that the disk is younger than the halo, and the apparent difference in age inferred from the nuclear chronometers and the globular clusters on one hand (although the last estimates are lower) and the luminosity function of white dwarfs on the other, can give us an indication of the delay between the formation of the halo and the thin-disk. As we have seen in chapter 6, successful chemical evolution models suggest that the appearance of the thin-disk may be delayed by several Gyr relative to the halo and the bulge.

In Figure 7.1 we report a result from chemo-dynamical models showing the appearance of the thin-disk only after 6 Gyr from the beginning of star formation in our Galaxy. The reason for such a delay in this chemo-dynamical approach arises from the fact that after the heating due to the first SNe the gas needs time to cool, before settling into the thin disk. Models of chemical evolution taking into account the dynamical effects through the infall also predict a long timescale for the formation of the local disk mainly to reproduce the G-dwarf metallicity distribution, as shown in chapter 6.

Finally, another way of deriving the age of the Galaxy is through the Hubble constant, H_o. Unfortunately, the broad range of inferred values of H_o^{-1} represents an uncertainty of a factor at least 1.5.

Table 7.3.   Galactic ages from other methods

Method	Age (Gyr)	Reference
Globular Clusters	$t_G = 11.8^{+2.1}_{-2.5}$	Gratton et al. (1997b)
Globular Clusters	$t_G \sim 12$	Caloi et al. (1997)
Globular Clusters	$t_G \sim 14$	Pont et al. (1998)
White Dwarfs	$7 \leq t_D \leq 11$	Iben & Laughlin (1989)
White Dwarfs	$8.0 \leq t_D \leq 10.5$	Wood (1990)
CMD of field stars	$t_D \sim 8$	Jimenez et al. (1998)
CMD of field stars	$t_D = 9 - 11$	Ng & Bertelli (1998)
Hubble law	$19 \leq H_o^{-1} \leq 30$	Sandage (1988)
Hubble law	$9 \leq H_o^{-1} \leq 11$	De Vaucouleurs et al. (1981)
Supernovae	$t_G = 14.2 \pm 1.7$	Perlmutter et al. (1999)
Planetary Nebulae	$10 \leq H_o^{-1} \leq 13.3$	Jacobi et al. (1990)
Open Clusters	$t_D \sim 8 - 9$	Carraro et al. (1999)

In Table 7.3 we report some determinations of the age of the Galaxy (t_G) from globular clusters, white dwarf luminosity functions, color-magnitude diagrams

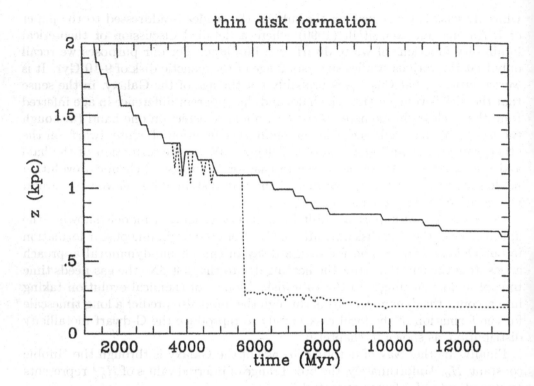

Figure 7.1. A result from the chemo-dynamical model of Burkert et al. (1992) illustrating the formation of the thin disk in the solar neighbourhood. The solid curve represents the outer edge of the old thin disk. It is evident that the emergence of a young thin disk occurs after 6 Gyr (dotted curve). From Burkert et al. 1992, Ap.J. Vol. 391, 651; reproduced here by kind permission of A. Burkert and the University of Chicago Press (copy right 1992).

(CMD) of field stars, open clusters and Hubble constant derived in turn from SNe, PNe and the Hubble law. Supernovae and planetary nebulae, in fact, can be used as standard candles to derive distances. Some of these methods (white dwarfs, color-magnitude diagram of field stars and open clusters) give only the age of the disk, t_D, at least the minimum age.

In summary, from the above discussion we cannot conclude that we know exactly the age of the Galaxy but we can indicate a possible range of 12-15 Gyr, with the age of the disk being in the range 8-10 Gyr.

Chapter 8

THE MILKY WAY AND OTHER SPIRAL
GALAXIES

In this chapter we compare the properties of the Milky Way with those of other spirals. We discuss the available observational information, in particular the chemical abundances and their correlations with galactic properties.

8.1 ABUNDANCES AND ABUNDANCE GRADIENTS

Giant HII regions are easily observable on the disks of nearly face-on spiral galaxies and these observations represent the usual current method of deriving information about the chemical composition of the ISM of external galaxies.

Abundances and abundance gradients derived from HII regions in external spirals have been reviewed by Edmunds (1989) and Diaz (1989). These data show that abundance gradients for O, N and S are present in disks of spirals. The gradients differ in size but they are not clearly correlated with mass or Hubble type of the galaxy. However, the absolute level of the abundances does show some systematic behaviour, for example O/H increases with the surface brightness and/or surface mass density. Diaz (1989) pointed out that N gradients in spirals are, in general, steeper than O gradients and that S gradients are somewhat flatter than those of oxygen which led some authors to ascribe this to the different stellar progenitors of O and S. Nucleosynthesis calculations indicate, in fact, that oxygen is produced by all stars from 10 to $50M_\odot$, whereas S originates from a more restricted mass range, namely from the most massive stars. However, recent gradients measured from the HII regions in the Milky Way do not show appreciable differences between the gradient of sulphur and that of oxygen (see Table 1.3). Vila-Costas and Edmunds (1992, 1993 and references therein) confirmed the existence of abundance gradients of N, O and S in external galaxies and found some correlations between the central abundances of spirals and their mass. They also indicated that barred spirals seem to have shallow gradients and non-barred spirals show a correlation of gradient slope with morphological type. The previously found correlation between abundances and surface mass

249

density was also confirmed. Henry et al. (1994) measured abundances in 10 HII regions in a Virgo Cluster spiral (NGC4254) and determined the gradients for O, N and S. The conclusion was that oxygen and sulphur gradients are similar and they are both flatter than the N gradient. Henry and Howard (1995) determined the abundance gradients in three nearby spirals, M33, M81 and M101. In particular, they measured the gradients of O, N and S and concluded that they are best represented by exponential functions and that the slopes of these abundance gradients remain constant across the disks. This is at variance with earlier suggestions (e.g. Diaz, 1989) indicating that there is a change in slope in disks; the same has been claimed for the disk of the Milky Way but not confirmed by all the studies.

One of the largest data sample for gradients in external spirals is from Zaritsky et al. (1994) who measured the gradient of (O/H) in 39 disk galaxies. Their study indicates that there are large differences among gradients in spirals, some have strong gradients, others have gradients similar to the Milky Way and others have almost no gradient. They also discussed whether abundance gradients are exponential and suggested that firm conclusions on the shape of abundance gradients await studies utilizing many observed regions per galaxy. They confirmed the tendency for barred-spirals to have shallower gradients than unbarred galaxies. This can perhaps be explained by the strong dynamical action exerted by the bar inducing large scale mixing processes in the disk (Martin and Roy, 1994).

In Figure 8.1 the (O/H) abundance gradients found in nearby spirals are shown. The abundance gradients are measured from HII regions and plotted as functions of the galactocentric distance normalized by the isophotal radius (R_{25}). They are expressed in units of dex/R_{25}. The choice of a normalization is fundamental in a comparison of abundance properties of different galaxies. In fact, adopting an absolute radial scale (dex/kpc) introduces variations among abundance properties which depend directly on the disk size. The figure shows that negative oxygen abundance gradients are found in disks of external spirals and that they are similar to the gradient found in the Milky Way disk. This is evident in Figure 8.2 where we show a comparison between the oxygen abundance gradient measured in the Milky Way, in NGC 628 and in M33. As one can see from this figure it is evident that the two external spirals resemble our Galaxy in possessing similar negative abundance gradients of oxygen. Therefore, abundance gradients are a common characteristic of spiral disks and must be linked to the mechanism of galaxy formation, as already discussed in chapter 6.

For the same spirals of figure 8.1 the N gradient has also been measured, so it is possible to derive the gradient in the (N/O) ratio. This ratio is important since it involves two chemical elements formed in different stellar mass ranges. Therefore it can be used either as a cosmic clock or as a constraint for theories of stellar nucleosynthesis. As we have seen in the previous chapters, N is be-

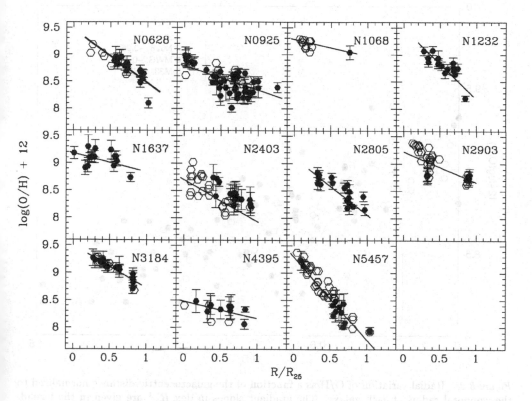

Figure 8.1. Oxygen abundance gradients in nearby spirals. The filled symbols represent data from van Zee (1998); the open symbols are data from various sources in the literature: NGC 0628- McCall et al. (1985); NGC 0925 - Zaritsky et al. (1994); NGC 1068- Evans and Dopita (1987), Oey and Kennicutt (1993); NGC 2403 - McCall et al. (1985), Fierro et al. (1986), Garnett et al. (1997); NGC2903- McCall et al. (1985), Zaritsky et al. (1994); NGC4395- McCall et al. (1985); NGC5457 - Kennicutt and Garnett (1996). The solid lines are the oxygen gradients derived from weighted least-squares fits (see Table 8.1). The oxygen abundances of the outermost HII regions in these galaxies are comparable to those found in dwarf galaxies. From van Zee 1998, in "Abundance Profiles: Diagnostic Tools for Galaxy History" ASP Conf. Series Vol. 147, 98; reproduced here by kind permission of L. van Zee and the Astronomical Society of the Pacific (copy right 1998).

lieved to be mostly a secondary element since it is a product of the CNO cycle and is formed at the expense of C and O already present in the star, although a primary N component originating in AGB stars is predicted by stellar nucleosynthesis studies. Massive stars should also produce N as a secondary element, although a primary N component originating from these stars would help in fitting the observed [N/Fe] versus [Fe/H] in the solar neighbourhood (see figure 6.14). From this plot, in fact, one can see that at low metallicities the [N/Fe] ra-

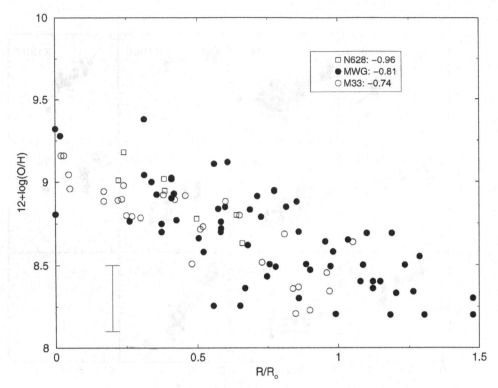

Figure 8.2. Radial variation of O/H as a function of the galactocentric distance normalized to the isophotal radius of each galaxy. The gradient slopes in dex R_o^{-1} are given in the legend. The data for the Milky Way are from Shaver et al. (1983), Afflerbach et al. (1997) and Vilchez and Esteban (1996). For NGC 628 and M33 the data are from Zaritsky et al. (1994). From Henry and Worthey 1999, P.A.S.P. Vol. 111, 919; reproduced here by kind permission of R. Henry and the University of Chicago Press (copy right 1999).

tio tends to flatten, thus indicating the necessity of having primary nitrogen from massive stars, as originally suggested by Matteucci (1986) and more recently by Maeder (1999) as a result of rotational diffusion. On the other hand, oxygen is a typical primary element formed at the expense of the original H and He. As we have seen in chapter 5, a secondary element is expected to show a different evolutionary behaviour than a primary one; in particular, the abundance of a purely secondary element would increase proportionally to the square of the abundance of its progenitor. However, this result is based on the Simple Model of chemical evolution which does not take into account the stellar lifetimes. When the stellar lifetimes are properly accounted for there is also the time-delay effect to be considered; in the particular case of N and O the effect is between a primary element, such as oxygen, which is produced on short timescales from massive stars, and a secondary element such as nitrogen, which is mostly produced on longer time scales from low and intermediate mass stars. As a consequence of this, the reader should beware of easy conclusions based only on the analytic solutions of the Simple Model which includes I.R.A. As shown in chapter 6 for the Milky Way, consideration of these effects (secondary and primary nature of N as well as relaxation of I.R.A.) leads to the prediction of a N/O gradient in galactic disks flatter than the oxygen gradient, in agreement with observations, thus arguing in favor of the fact that some primary N is indeed required. On the other hand, the predictions of the Simple Model for a purely secondary element would indicate a N/O gradient as steep as that of O/H. For external galaxies there is also evidence for a flat negative gradient of N/O along the disks and therefore we are inclined to conclude that also in external spirals there is the need for some primary nitrogen. In Table 8.1 are shown the gradients of O/H and N/O derived by van Zee (1998) from optical spectroscopy of 180 HII regions in 11 nearby spiral galaxies.

Table 8.1.   Radial abundance gradients in spiral galaxies

Object	$R_{25}[arcsec]$	Gradient $((O/H)[dex/R_{25}])$	Gradient $((N/O)[dex/R_{25}])$
NGC 0628	314.	-0.99 ± 0.14	-0.57 ± 0.12
NGC 0925	314.	-0.45 ± 0.08	-0.34 ± 0.04
NGC 1068	212.	-0.30 ± 0.07	-0.01 ± 0.09
NGC 1232	222.	-1.31 ± 0.20	-0.32 ± 0.10
NGC 1637	120.	-0.37 ± 0.14	-0.30 ± 0.17
NGC 2403	656.	-0.77 ± 0.14	-0.40 ± 0.08
NGC 2805	189.	-1.05 ± 0.17	-0.29 ± 0.06
NGC 2903	378.	-0.56 ± 0.09	-0.57 ± 0.17
NGC 3184	222.	-0.78 ± 0.07	-0.77 ± 0.12
NGC 4395	395.	-0.32 ± 0.19	-0.04 ± 0.11
NGC 5457	865.	-1.52 ± 0.09	-0.65 ± 0.09

In the first column of Table 8.1 are shown the galaxies, in the second column the isophotal radius, and in the third and fourth column the O/H and N/O gradients, respectively.

Finally, a study of the metallicity in the gas in the extreme outer regions of disks of spirals, up to $\sim 1.5 - 2\,R_{25}$ and sometimes further (Ferguson 1999) has revealed that there is no evidence for a flattening of the gradients and that the chemical abundances (N and O) are $\sim 10-15\%$ solar. These gradients are shown in Figure 8.3 where the radial variations of O/H and N/O as functions of R/R_{25} are plotted.

Lastly, it may be noted that recently stellar oxygen gradients have been measured in external galaxies. For example Monteverde et al. (1997) measured the stellar oxygen gradient in M33 from the analysis of 6 B-supergiants and found $\frac{d log(O/H)}{dR} = -0.16 \pm 0.06$ dex kpc^{-1}, while Venn et al. (1998) derived stellar gradients for oxygen, other α-elements and s-process elements in the disk of M31 from the analysis of A-type supergiants and concluded that the oxygen stellar gradient agrees within the uncertainties with the nebular one, which is of the order of -0.03 dex kpc^{-1} in the galactocentric distance range 5-20 kpc. However, given the paucity of stars used for the previous studies it seems still premature to draw firm conclusions, even though the study of stellar abundances in external galaxies is a very promising one and will be of fundamental importance to study the chemical evolution of Local Group Galaxies.

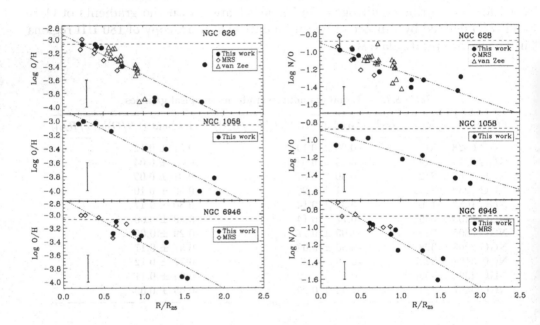

Figure 8.3. Radial variation of O/H (left panel) and N/O (right panel) as a function of galactocentric distance, expressed in terms of optical radius. The inner points are derived from the measurements of McCall et al. (1985) and are indicated by filled circles. The horizontal dashed line indicates the solar abundance. From Ferguson 1999, in "Chemical Evolution from Zero to High Redshift" eds. J.R. Walsh and M.R. Rosa, p. 109; reproduced here by kind permission of A. Ferguson and Springer Verlag (copy right 1999).

8.2 CORRELATIONS BETWEEN ABUNDANCES AND GALAXY PROPERTIES

We summarize here the main observed correlations among observables in disks of external spirals:

- *Metallicity versus galaxy luminosity/mass*

The correlations between the abundance of oxygen at one disk scale length from the galaxy nucleus with the maximum rotational velocity and luminosity are displayed in Figure 8.4. From this figure we can see that, although the maximum rotational velocity is probably a better measure of the mass than the galactic magnitude M_B, there is no qualitative difference between the two plots. The meaning of the mass-metallicity relation is not yet completely understood, especially in spiral galaxies. On the other hand, a similar correlation exists in elliptical galaxies and is generally explained as being due to galactic winds which occur later in massive than in small ellipticals. This means that a massive galaxy continues to form stars for a longer period resulting in a higher average metallicity than would be achieved in a smaller object. In fact, the idea is that after the occurrence of the galactic wind effectively no star formation will take place. This interpretation of the mass-metallicity relation in ellipticals was first suggested by Larson (1974). However, galactic winds do not seem to be playing an important role in the Milky Way, and therefore presumably they are not very important in external spirals. Moreover, even for elliptical galaxies it is not certain that the Larson picture is correct, since there are observational indications that the [Mg/Fe] ratio in the dominant stellar population in the nuclei of ellipticals, increases with galactic luminosity, thus implying, as a possible explanation, a shorter duration of the star formation period in massive ellipticals, within the framework of the time-delay model (Matteucci, 1994 and references therein). Actually, the same interpretation might possibly be applied also to spiral galaxies; in particular, Prantzos and Boissier (2000), by means of a chemical evolution model devised for the Milky Way and of the type discussed in chapter 6, suggested that, in order to reproduce the majority of the observables for a sample of spirals, the timescale for star formation should be longer in low mass disk spirals than in more massive ones.

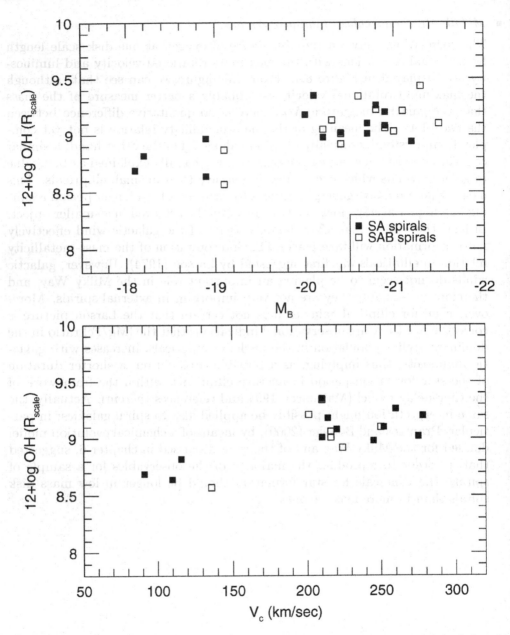

Figure 8.4. Top: the correlation of spiral galaxy abundance (O/H) at one disk scale-length from the galaxy nucleus vs. galaxy luminosity. Bottom: abundance vs. maximum rotational velocity V_C. From Garnett 1998, in "Abundance Profiles: Diagnostic Tools for Galaxy History" ASP Conf. Series Vol. 147, 78; reproduced here by kind permission of D. Garnett and the Astronomical Society of the Pacific (copy right 1998).

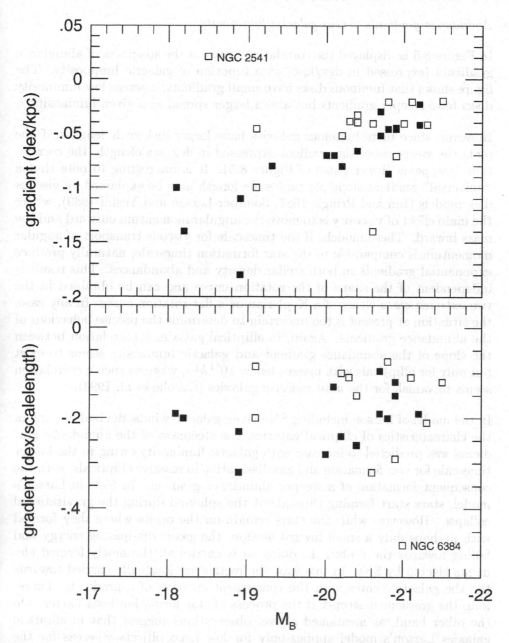

Figure 8.5. The correlation between the steepness of the oxygen gradient vs. M_B. The upper panel shows the gradient expressed in dex/kpc and the lower panel the gradient expressed in dex/scalelength. From Garnett et al. 1997, Ap.J. Vol. 489, 63; reproduced here by kind permission of D. Garnett and the University of Chicago Press (copy right 1997).

- *Abundance gradients versus galactic luminosity*

In Figure 8.5 is displayed the correlation between the steepness of abundance gradients (expressed in dex/kpc) as a function of galactic luminosity. The figure shows that luminous disks have small gradients, whereas low luminosity disks have steeper gradients but also a larger spread at a given luminosity.

However, since more luminous galaxies have larger disk scale lengths, if one plots the steepness of the gradient expressed in dex/scalelength, the correlation disappears (lower panel of Figure 8.5). It is interesting to note that a "universal" gradient slope per unit scale length may be explained by viscous disk models (Lin and Pringle 1987, Sommer-Larsen and Yoshii 1989), where the main effect of viscosity is to move the angular momentum outward and the mass inward. These models, if the timescale for viscous transport of angular momentum is comparable to the star formation timescale, naturally produce exponential gradients in both stellar density and abundances. This result is independent of the shape of the rotation curve and can be obtained in the two extreme cases either of a Keplerian or a flat rotation curve. In any case, the situation at present is too uncertain to determine the precise behaviour of the abundance gradients. Again, in elliptical galaxies a correlation between the slope of the abundance gradient and galactic luminosity seems to exist but only for ellipticals with masses below $10^{11} M_\odot$, whereas such a correlation seems to vanish for the most massive galaxies (Carollo et al. 1993).

In the model of Larson including SN-driven galactic winds, devised to explain the characteristics of elliptical galaxies, the steepness of the abundance gradients was predicted to increase with galactic luminosity owing to the longer timescale for star formation and gas dissipation in massive ellipticals, with the consequent formation of a steeper abundance gradient. In fact, in Larson's model, stars start forming throughout the spheroid during the gravitational collapse. However, while the stars remain on the orbits where they formed with perhaps only a small inward motion, the gas is dissipating energy and falling towards the center. In doing so it carries all the newly formed elements ejected by SNe. In this way, the metals are gradually carried towards the the galactic center with the consequent creation of a gradient. Therefore, the gradient is steeper if the process of star formation lasts longer. On the other hand, as mentioned above, observations suggest that in elliptical galaxies Larson's model applies only for low mass objects whereas for the more massive ellipticals one should invoke a different mechanism of galaxy formation, such as perhaps mergers of gaseous lumps. Whether a model with SN-driven winds is suitable for spiral galaxies is not clear in the light of our present knowledge, as already discussed. Such a mechanism would produce a positive gradient in the [α/Fe] ratio as a function of galactocentric distance, since it would predict a decreasing duration of the star formation period with increasing galactocentric radius, the contrary of the inside-out picture.

- *Metallicity versus surface brightness*

The uniformity of abundance gradients as a function of scale length, if true, suggests that there should also be a correlation between metallicity and surface brightness, as indeed noted by McCall (1982) and Edmunds and Pagel (1984) for late-type spirals. The existence of such a correlation has partly served as a basis for theories of self-regulated star formation, in which the energy produced by young stars and SNe feeds back into the ISM and inhibits further star formation (see chapter 3).

In Figure 8.6 is shown the correlation between the oxygen abundance, as measured from HII regions in the disk of several spirals, and the surface brightness. The plot indicates the existence of a local metallicity-surface brightness relationship in galactic disks.

However, early-type spirals do not follow the same correlation as the late-type spirals. In Figure 8.7 is shown the characteristic metallicity at two fixed values of disk surface brightness for a sample of spirals having either I- or R-band surface photometry. The figure shows that there is a remarkable correlation between the metallicity at a given surface brightness and the luminosity of a galaxy, in such a way that luminous galaxies are more metal-rich than low luminosity galaxies at the same surface brightness. As a consequence of this, it seems that the metallicity in spirals is connected both to the stellar mass density profile and to the total galaxy mass. In other words, this can suggest that *the star formation rate is connected to both the surface gas density and the total surface mass density*, a correlation adopted by the most successful chemical evolution models discussed in the previous chapters.

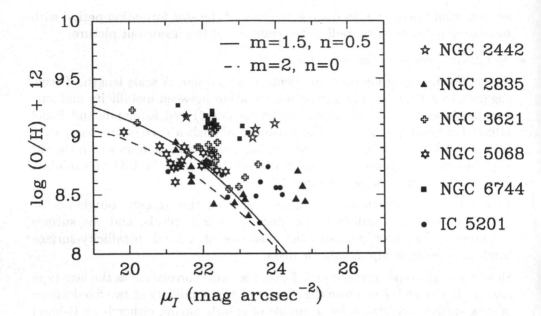

Figure 8.6. The observed relationship between the oxygen abundance in HII regions of several spirals and the underlying surface brightness (symbols), together with the predictions of models adopting the star formation rate of Dopita and Ryder (1994) with indices m and n as defined in eq. 3.38. Figure from Ryder 1998, in "Abundance Profiles: Diagnostic Tools for Galaxy History" ASP Conf. Series Vol. 147, 83; reproduced here by kind permission of S. Ryder and the Astronomical Society of the Pacific (copy right 1998).

- *Star formation rate versus gas density*

As already discussed in Section 3.3, there is a clear indication of a correlation between the SFR and the total surface gas density ($HI + H_2$) in spiral galaxies. In particular, Kennicutt (1998b) suggested a law of star formation depending on the total surface gas density such as in eq. (3.36) or alternatively in eq. (3.44). Both these laws provide a very good fit to the global SFR over a density range extending from the most metal-poor disks to the cores of the most luminous starburst galaxies (see figure 3.5).

No clear correlation exists between the H_2 and the SFR. This is illustrated in Figure 8.8 where the disk averaged SFR per unit area is plotted as a function of the average densities of HI and H_2. As discussed in chapter 3, this lack of correlation is very probably due to variations in the conversion factor CO/H_2, so firm conclusions cannot be drawn concerning this point. Figure 8.8 also shows that the more luminous metal-rich disks present a better defined SFR versus H_2 relation and is very similar to that observed for the SFR versus HI (always in the same figure). This may be due to the fact that the galactic

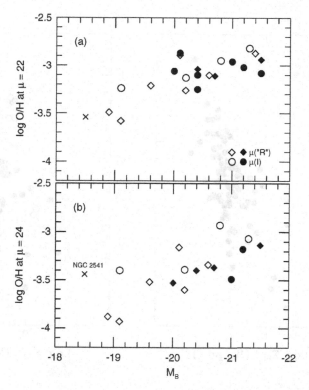

Figure 8.7. The abundance of O/H at a fixed value of galaxy surface brightness vs. M_B. Top: abundances at $22 magarcsec^{-2}$. Bottom: abundances at $24 magarcsec^{-2}$. More luminous spirals have higher oxygen abundances at a fixed surface brightness. Open symbols from Garnett et al. (1997). From Garnett 1998, in "Abundance Profiles: Diagnostic Tools for Galaxy History" ASP Conf. Series Vol. 147, 98; reproduced here by kind permission of D. Garnett and the Astronomical Society of the Pacific (copy right 1998).

CO/H_2 conversion factor, valid for regions of near solar metallicity, fails if applied to regions of much lower metallicity.

8.3   FIELD AND CLUSTER SPIRALS

The distribution of the total gas (HI $+ H_2$) is known only in a few cases for external spirals owing to the difficulty of measuring the molecular hydrogen (see chapter 4); but the HI distribution along the disks of spirals is better known. In figure 8.9 we report the observed radial profiles of the HI in spirals of the Virgo cluster compared to the same profiles for field spirals. In particular, the points refer to the azimuthally averaged HI surface densities, while the solid lines represent the mean distributions for field spirals of the same morphological type. The galaxies appear divided into three groups: those with HI deficiencies, the intermediate cases and those with no deficiencies. In the lower panel of figure 8.9

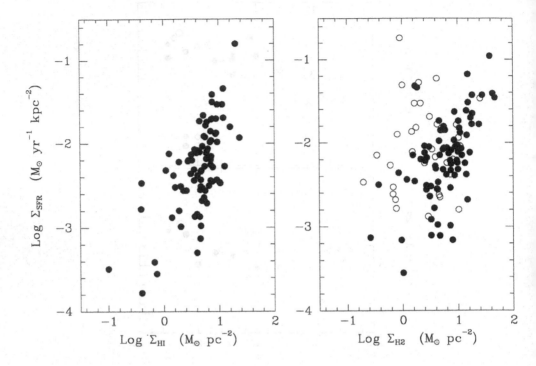

Figure 8.8. The correlation of the disk-averaged SFR per unit area with the average surface densities of HI (left) and H_2 (right). The H_2 densities were derived by using a constant CO/H_2 conversion factor. In the right panel, the filled circles denote galaxies with $L_B > 10^{10} L_\odot$, while the open circles denote galaxies with $L_B < 10^{10} L_\odot$. From Kennicutt 1998, Ap.J. Vol. 498, 541; reproduced here by kind permission of R. Kennicutt and the University of Chicago Press (copy right 1998).

are shown the abundance gradients; again the points refer to the Virgo spirals and the solid lines to the average gradient for field spirals of the same morphological type.

This figure is instructive because it suggests the following considerations:

- the most HI deficient galaxies are the central ones in Virgo and show little evidence for abundance gradients but high abundances, which are larger than those in the spirals in the periphery of the cluster (0.3 to 0.5 dex in O/H).

- The HI deficient spirals in Virgo have larger mean abundances than field galaxies of comparable luminosity or Hubble type, whereas the spirals at the periphery of the cluster are indistinguishable from the field spirals.

These facts may be explained in the framework of spiral galaxies falling into the Virgo cluster, where the HI deficiency can be due to the removal of gas from the infalling spirals by the cluster environment (ram pressure stripping). Chemical evolution models devised to explain this effect have been presented by Kennicutt et al. (1994) and suggested that is indeed possible to explain the HI deficiency and the high metal abundance in the core spirals by curtailment of infall of metal-poor gas in these objects, as opposed to normal field spirals or peripheral Virgo spirals where the metal-poor infall is a continuous process. However, much work still needs to be done to understand the evolution of spirals and the differences between cluster and field spirals.

8.4 FORMATION AND EVOLUTION OF GALACTIC DISKS

Chemical evolution models for external spirals have been presented by several authors (e.g. Diaz and Tosi 1984; Molla et al. 1996; Prantzos and Boissier 2000). In order to build a reliable chemical evolution model for external spirals one needs to know at least the radial abundance gradients and the radial gas distribution and, if possible, the SFR distribution. Among the most studied spirals are M31, M33 and M101 for which these data are now available. Generally, the most important parameters required to study the evolution of galactic disks are the SFR and the IR, in other words the mechanism of star formation and the history of formation of the disks (inflows and/or outflows).

For the Milky Way we have seen in chapter 6 that the best chemical evolution model suggests that the galactic disk formed mostly out of extragalactic gas and in an inside-out fashion with the innermost regions forming faster than the outermost ones. Since the radial properties of external spirals are similar to those of the Milky Way one must assume that the inside-out mechanism is common to all spirals and that the differences among abundance gradients derive mainly from the different distributions of the total mass and the gas mass. In this way, one may think of modelling the evolution of external spirals just by adopting some "scaling laws" calibrated on the Milky Way.

Evidence for inside-out formation of galactic disks comes also from the analysis of the colour gradient in M31 (Josey and Arimoto 1992) and in the the disk of the Milky Way (Boissier and Prantzos 1999). In these models the disk is assumed to have formed by an inside-out mechanism and the bias in the infall (namely a faster infall rate in the innermost regions) produces a variation in the predicted stellar populations which well reproduces the observed colour gradients. On the other hand, models with a constant timescale for the infall along the disk do not fit the observations. Therefore, the chemical and photometric information suggests that the disk of the Milky Way as well as the disks of other spirals must have formed over a long period of time although the length of this period may vary from galaxy to galaxy. This is a very important conclusion in the context of galaxy formation theories.

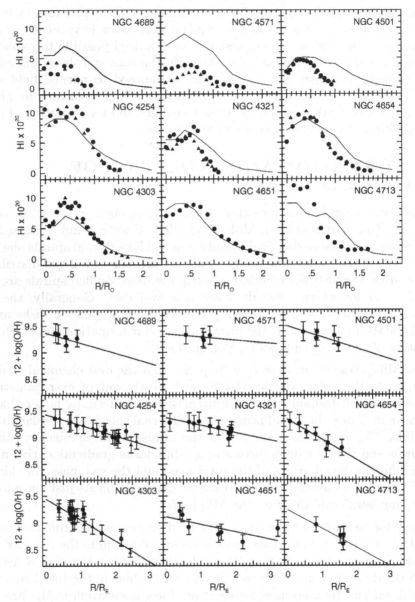

Figure 8.9. A comparison between cluster and field spirals. In the top panel the HI profiles as functions of the galactocentric distance for the Virgo spirals (filled circles from Warmels (1988) and filled triangles from Cayatte et al. (1990) and for the case of NGC 4571 from van der Hulst et al. (1987), and for the field spirals (solid lines). The top row shows the HI-deficient (cluster core) galaxies and the bottom row shows the HI-normal (peripheral) galaxies. The individual points represent the radially averaged HI column densities, whereas the solid lines represent average radial profiles for the corresponding Hubble type. In the lower panel are plotted the oxygen abundances as functions of the galactocentric distance normalized to the effective radius. The solid lines refer to the gradients measured by Zaritsky et al. (1994). Figure from Skillman et al. 1996, Ap.J. Vol. 462, 147; reproduced here by kind permission of E. Skillman and the University of Chicago Press (copy right 1996).

Standard hierarchical models of galaxy formation also seem to require a late formation of massive galactic disks (e.g. Mo, Mao and White, 1998 and references therein). In these models, dissipationless dark matter aggregates into larger and larger clumps as gravitational instability amplifies the weak density perturbations produced at early times. The gas, associated with the dark halos formed in the above mentioned way, cools and condenses within them eventually forming the galaxies we see today. After cooling, in fact, the gas produces an independent self-gravitating unit which can form stars that heat and enrich the rest of the gas, perhaps even producing a wind ejecting the gas outside the halo. Here is the link between the cosmological galaxy formation and the evolution of gas and stars that we have described in this book: on the one hand, the gravitational processes which determine the formation, the abundance, the structure and the kinematics of dark halos are simulated either by N-body methods or semi-analytical models; on the other hand, the evolution of the gas requires the knowledge of the star formation process, the feedback between stars and gas and the stellar nucleosynthesis.

Although a real merger between the two approaches has not yet been accomplished, it is of considerable interest that they suggest a similar conclusion about the formation of galactic disks.

8.5   THE MILKY WAY AT HIGH REDSHIFT

We can conclude, from the previous sections, that the comparison between the properties of the Galaxy and those of external spirals strongly suggests that the Milky Way is a quite common spiral. For this reason, it is interesting to study the evolution of the galactic abundances as functions of redshift in order to put constraints on the early phases of the evolution of spiral galaxies.

In figure 8.10 we show the predictions of the Chiappini et al. (1999) model relative to the [O/Fe] ratio in the solar vicinity as a function of redshift for different cosmologies. In particular, a cosmology is identified by the age of the galaxy formation z_f, the density parameter Ω_o and the cosmological constant Λ. The interesting aspect of this figure is the sharp increase of the [O/Fe] ratio at very high redshift. The spike in [O/Fe] corresponds to the halo thick-disk phase and is due to the rapid increase of metallicity in the early phases of galactic evolution. Therefore, what we know as the [O/Fe] plateau, extending from [Fe/H]$\sim$-3.0 dex up to [Fe/H]=-1.0 dex, is confined in a very small redshift range , $\Delta z = 0.01$.

The same behaviour shown for the [O/Fe] versus redshift is expected for the [O/Zn] ratio, since Zn probably evolves in lockstep with Fe (see chapters 1 and 6); this is of interest in studying the DLA systems at high redshift. In fact, the abundances of α-elements and Zn have been measured in a large sample of DLA (Pettini et al. 1997;1999) since they are elements almost unaffected by dust. In this way, we can use predictions concerning our Galaxy to test whether these high redshift objects can be protospirals, as has often been suggested, or other

266

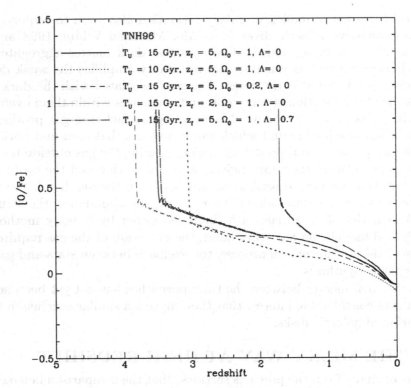

Figure 8.10. Predicted [O/Fe] versus redshift for different cosmologies as indicated in the figure (T_U is the age of the universe, z_f is the redshift of galaxy formation, Ω_o is the density parameter and Λ is the cosmological constant). The stellar yields from massive stars adopted in the model are from Thielemann et al. (1996).From Chiappini et al. 1999, Ap.J. Vol. 515, 226; reproduced here by kind permission of C. Chiappini and the University of Chicago Press (copy right 1999).

objects. As an example of this, we recall that these recent studies of DLA systems indicate that the metallicity of these high-redshift and low metallicity objects ($z < 1.5$ and $[< Zn/H >]$=-1.03±0.3 dex) seems not to evolve with redshift (see figure 8.11) and that the [Si/Zn] ratios are roughly solar. A solar [α/Fe] ratio at a metallicity of [Fe/H] $\sim$ -1.0 dex implies, according to the time-delay model discussed in chapter 6, that what we observe are objects where star formation proceeded very slowly, thus allowing a large Fe (and iron-peak element) pollution from type Ia SNe even for low absolute values of [Fe/H]. The situation is perfectly illustrated in figure 6.3, showing the different [O/Fe] versus [Fe/H] relations arising from different star formation histories. However, there is a warning about using Si as a typical α-element since this element is produced in a non-negligible way also in SN Ia (see Table 2.1). Therefore, Si is not expected to show a large overabundance relative to iron at low metallicity. A more secure test for the lack of overabundance in α-elements in DLA systems would be provided by the abundances of other α-elements mainly produced by massive stars, such as O, Mg and S. In conclusion, the lack of an overabundance of α elements in the DLA systems, if true, indicate that they could be either the external regions of galactic disks or magellanic irregular small galaxies, and that they certainly do not trace the galaxy population responsible for the bulk of star formation at high redshift.

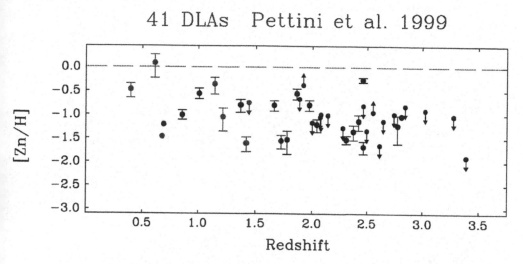

Figure 8.11. The abundance of Zn versus redshift for a sample of DLAs consisting of 34 systems in the study of Pettini et al. (1997) plus the six systems at $z < 1.5$ measured by Pettini et al. (1999). The solar abundance [Zn/H]=0.0 is shown by the broken line. From Pettini et al. 1999, Ap.J. Vol. 510, 576; reproduced here by kind permission of M. Pettini and the University of Chicago Press (copy right 1999).

In conclusion, it is worth stressing again that abundance ratios represent a very strong constraint for inferring the nature of high redshift objects in general, and that the study of the chemical evolution of the Galaxy is a necessary step in the process of understanding the evolution of all galaxies and the universe.

Chapter 9

REFERENCES

Abel, T., Anninos, P., Norman, M.L.,Zhang, Y., 1998, ApJ **508**, 518

Abia, C. Boffin, H.M.J., Isern, J., Rebolo, R., 1993, A&A **272**, 455

Afflerbach, A., Churchwell, E., Werner, M.W., 1997, ApJ **478**, 190

Allen, C.W., 1954, MNRAS **114**, 387

Anders, E., Grevesse, N., 1989, Geochim. Cosmochim. Acta **53**, 197

Andersson, H., Edvardsson, B., 1994, A&A **290**, 590

Appenzeller, I. Fricke, K.J., 1972, A&A **18**, 10

Appenzeller, I., Tscharnuter, W., 1973, A&A **25**, 125

Arimoto, N., Yoshii, Y., 1987, A&A **173**, 23

Arnett, D.W., 1991, in " Frontiers of Stellar Evolution", ed. D.L. Lambert, ASP Conf. Series Vol. 20, p. 389 (A91)

Arnett, D.W., 1995, ARA&A **33**, 115 (A95)

Armoski, B.J., Sneden, C., Langer, G.E., Kraft, R.P., 1994, A.J., **108**, 1364

Bahcall, J.N., 1984, ApJ **276**, 169

Barbuy, B., 1999, ApSS **265**, 319

Barbuy, B., 1988, A&A **191**, 121

Barbuy, B., Erdelyi-Mendes, M., 1989 A&A **214**, 239

Barbuy, B., Grenon, M., 1990, in "Bulges of Galaxies", ESO/CTIO Workshop, eds B.J. Jarvis and D.M. Terndrup, p. 83

Becker, S.A., Iben, I. Jr, 1979, ApJ **232**, 831

Becker, S.A., Iben, I. Jr, 1980, ApJ **237**, 111

Beer, H., 1990, in "Astrophysical Ages and Dating Methods", Gif sur Yvette: Edition Frontieres

Beers, T.C., Sommer-Larsen, J., 1995, ApJ Suppl. **96**, 175

Berman, B.C., Suchov, A.A., 1991, ApSS **184**, 169 (BS)

Bertelli, G., Bressan, A., Chiosi, C., 1985, A&A **150**, 33

Bessel, M.S., Sutherland, R.S., Ruan, K., 1991, ApJ **383**, L71

Bethe, H.A., Wilson, J.R., 1985, ApJ **295**, 14

Bhat, C.L., Houston, B.P., Issa, M.R., Mayer, C.J., Wolfendale, A.W., 1984 in "Gas in the Interstellar Medium", ed. P.M. Gondhalekar, p. 39, RAL

Binney, J.J., Davies, R.L., Illingworth, G.D., 1990, ApJ **361**, 78

Blitz, L., Spergel, D., Teuben, P.J., Hartmann, D., Burton, W. B., 1999, ApJ **514**, 418

Blöcker, T., Schönberner, D., 1991, A&A **244**, L43

Boffin, H.M.J., Paulus, G., Arnould, M., Mowlavi, N., 1993 A&A **279**, 173

Bolte, M., Hogan, C.J., 1995, Nature **376**, 399

Boissier, S., Prantzos, N., 1999, MNRAS **307**, 857

Bond, H.E., 1970, ApJ Suppl. **22**, 117

Bonifacio, P., Molaro, P., 1997, MNRAS **285**, 847

Branch, D., Nomoto, K., 1986, A&A **164**, L13

Braun, R., Burton, W.B., 1999, A&A **341**, 437

Bressan, G.A., Bertelli, G., Chiosi, C., 1981, A&A **102**, 25

Buonanno, R., Corsi, C.E., Fusi Pecci, F., Fahlman, G.G., Richer, H.B., 1994 ApJ Lett. **430**, 121

Burkert, A., Hensler, G., 1987, MNRAS **225**, 21P

Burkert, A., Hensler, G., 1988, A&A **199**, 131

Burkert, A., Hensler, G., 1989 in "Evolutionary Phenomena in Galaxies" eds. J.E. Beckman and B.E.J. Pagel, Cambridge Univ. Press p. 230

Burkert, A., Truran, J.W., Hensler, G., 1992, ApJ **391**, 651

Burles, C., Tytler, D., 1998, Space Science Rev. **84**, 65

Burton, W.B., Gordon, M.A., 1978, A&A **63**, 7

Butcher, H.R., 1987, Nature **328**, 127

Caloi, V., D'Antona, F., Mazzitelli, I., 1997, A&A **320**, 823

Cameron, A.G.W., 1982, in "Essays in Nuclear Astrophysics", ed. C.A. Barnes et al. Cambridge Univ. Press, p. 23

Cameron, A.G.W., Fowler, W.A., 1971, ApJ **164**, 111

Capaccioli, M., della Valle, M., Rosino, L., D'Onofrio, M., 1989, A.J. **97**, 1622

Cappellaro, E., Turatto, M., Tsvetkov, D.Yu., Bartunov, O.S., Makarova, I.N., 1993, A&A **273**, 383

Cappellaro, E., Turatto, M., Tsvetkov, D.Yu., Bartunov, O.S., Pollas, C., Evans, R., Hamuy, M., 1997, A&A **322**, 431

Carbon, D.F., Barbuy, B., Kraft, R.P., Friel, E.D., Suntzeff, N.B., 1987, PASP **99**, 335

Carigi, L., 1994, ApJ **424**, 181

Carigi, L., 1996, Rev. MexAA **32**, 179

Carlberg, R.G., Dawson, P.C., Hsu, T., Vandenberg, D.A., 1985, ApJ **294**, 674

Carney, B.W., 1996, P.A.S.P. **108**, 900

Carollo, C.M., Danziger, I.J., Buson, L., 1993, MNRAS **265**, 553

Carollo, C.M., de Zeeuw, P.T., van der Marel, R.P., Danziger, I.J., Qian, E.E., 1995, ApJ Lett. **441**, L25

Carraro, G., Ng, Y.K., Portinari, L., 1998a, MNRAS **296**, 1045

Carraro, G., Lia, C., Chiosi, C., 1998b, MNRAS **297**, 1021

Carraro, G., Vallenari, A., Girardi, L., Richichi, A., 1999, A&A **343**, 825

Carretta, E., Gratton, R.G., 1997, A&A Suppl. **121**, 95

Castro, S., Rich, R.M., Grenon, M., Barbuy, B., McCarthy, J.K., 1997, A.J. **114**, 376

Cayatte, V., Balkowski, C., van Gorkom, J.H., Kotanyi, C., 1990, A.J. **100**, 604

Chiappini, C., Matteucci, F., Gratton, R. 1997, ApJ **477**, 765

Chiappini, C., Matteucci, F., Beers, T.C., Nomoto, K., 1999, ApJ **515** , 226

Chiosi, C., Caimmi, R., 1979, A&A **80**, 234

Chiosi, C., 1987, in "Nucleosynthesis and Stellar Evolution", 16th Saas-Fee Course, ed. B. Hauck et al., Geneva Observatory, p. 199

Chiosi, C., Nasi, E., Sreenivasan, S.R., 1978, A&A **63**, 103

Chiosi, C., Nasi, E., Bertelli, G., 1979, A&A **74**, 62

Chiosi, C., 1980, A&A **83**, 206

Chiosi, C., Matteucci, F., 1980, Mem. Soc. Astron. It. **51**, 107

Clarke, C.J., 1989, MNRAS **238**, 283

Clayton, D.D., 1985, in "Nucleosynthesis: Challenges and New Developments", eds. D.W. Arnett and J.W. Truran, Univ. of Chicago Press, p. 65

Christlieb, N., Bessell, M.S., Beers, T.C., et al., 2002, Nature, **419**, 904

Clayton, D.D., 1987, ApJ **315**, 451

Clayton, D.D., 1988, MNRAS **234**, 1

Copi, C. J., 1997, ApJ **487**, 704

Cowan, J.J., Thielemann, F.K., Truran, J.W., 1991, ARA&A **29**, 447

Cowan, J.J., Thielemann, F.K., Truran, J.W., 1987, ApJ **323**, 543

Crovisier, J., 1978, A&A **70**, 43

Dahmen, G., Wilson, T.L., Matteucci, F., 1995, A&A **295**, 194

Dame, T.M., 1993 in "Back to the Galaxy", AIP Publ., eds. S.S. Holt and F. Verter, p. 267

Danly, L., 1989, ApJ **342**, 785

D'Antona, F., Mazzitelli, I., 1996, ApJ **456**, 329

D'Antona, F., Mazzitelli, I., 1990, ARA&A **28**, 139

D'Antona, F., Mazzitelli, I., 1984, A&A **138**, 431

D'Antona, F., Matteucci, F., 1991, A&A **248**, 62

D'Antona, F., 1995, in "The Bottom of the Main Sequence - And Beyond" ed. C.G. Tinney, Springer, p. 13

Danziger, I.J., 1999, in "The Chemical Evolution of the Milky Way: Stars versus Clusters", ed.s. F. Matteucci and F. Giovannelli, Kluwer (Dordrecht), ASSL Vol. 255, p. 59

Dearborn, D.S.P., Steigman, G., Tosi, M., 1996, ApJ **465**, 877

De Boer, K.S., Savage, B.D., 1984, A&A **136**, L7

Della Valle, M., Livio, M., 1994, A&A **286**, 786

272

D'Ercole, A., Brighenti, F., 1999, MNRAS **309**, 941

De Vaucouleurs, G., Peters, W.L., Bottinelli, L., Gouguenheim, L., Patural, G., 1981, ApJ **248**, 408

DeYoung, D.S., Gallagher, J.S. III, 1990, ApJ **356**, L15

Diaz, A.I., 1989, in "Evolutionary Phenomena in Galaxies", eds. J.E. Beckman and B.E.J. Pagel, Cambridge Univ. Press p. 377

Diaz, A., Tosi, M., 1984, MNRAS **208**, 365

Donas, J., Deharveng, J.M., Laget, M., Milliard, B., Huguenin, D., 1987, A&A **180**, 12

Dopita, M.A., Ryder, S.D., 1994, ApJ **430**, 163

Duncan, D.K., Rebull, L.M., 1996, PASP **108**, 738

Edmunds, M.G., 1975, ApSS **32**, 483

Edmunds, M.G., 1989, in "Evolutionary Phenomena in Galaxies", eds. J.E. Beckman and B.E.J. Pagel, Cambridge Univ. Press p. 356

Edmunds, M.G., 1990, MNRAS **246**, 678

Edmunds, M.G., 1992, in "Elements in the Cosmos" eds. M.G. Edmunds and R.J. Terlevich, Cambridge Univ. Press p. 289

Edmunds, M.G., Greenhow, R.M., 1995, MNRAS **272**, 241

Edmunds, M.G., Greenhow, R.M., Johnson, D., Klueckers V., Vila, M.B., 1991, MNRAS **251**, 33P

Edmunds, M.G., Pagel, B.E.J., 1984, MNRAS **211**, 507

Edvardsson, B., Andersen, J., Gustafsson, B., Lambert, D.L., Nissen, P.E., Tomkin, J., 1993, A&A **275**, 101

Eggen, O.J., Lynden-Bell, D., Sandage, A.R., 1962, ApJ **136**, 748 (ELS)

Elmegreen, B.G., 1999, ApJ **517**, 103

Esteban, C., Peimbert, M., 1995, A&A **300**, 78

Evans, I.N., Dopita, M.A., 1987, ApJ **319**, 662

Faber, S.M., Friel, E.D., Burstein, D., Gaskell, C.M., 1985, ApJ Suppl. **57**, 711

Feltzing, S., Gustafsson, B., 1998, A&A Suppl. **129**, 273

Ferguson, A., Wyse, R., Gallagher, J.S., 1999 in "Chemical Evolution from Zero to High Redshift", eds. J.R. Walsh and M.R. Rosa, Springer, p. 109

Ferrini, F., Molla, M., Pardi, M.C., Diaz, A.I., 1994, ApJ **427**, 745

Ferrini, F., Penco, U., Palla F., 1990, A&A **231**, 391

Fich, M., Tremaine, S., 1991, ARA&A **29**, 409

Fich, M., Silkey, M., 1991, ApJ **366**, 107

Fierro, J., Torres-Peimbert, S., Peimbert, M., 1986, PASP **98**, 1032

Filippenko, A.V., 1991, in "SN1987A and Other Supernovae" eds. I.J. Danziger and K. Kjär, ESO Publ., p. 343

Fitzsimmons, A., Dufton, P.L., Rolleston, W.R.J., 1992, MNRAS **259**, 489

Fowler, W.A., Meisl, C.C., 1986, in "Cosmogonical Processes" eds. W.D. Arnett et al. Singapore, VNU Press, p. 83

Fowler, W.A., 1987, Quart. J. RAS **28**, 87

Franco, J., Shore, S.N., 1984, ApJ **285**, 813

François, P., Matteucci, F., 1993, A&A **280**, 136

François, P., Vangioni-Flam, E., Audouze, J., 1990, ApJ **361**, 487

Friel, E.D., 1995, ARA&A **33**, 381

Friel, E.D., 1999, in "Galaxy Evolution: Connecting the distant Universe with the local fossil record", ApSS **265**, 271

Friel, E.D., Janes, K.A., 1993, A&A **267**, 75

Forestini, M., Charbonnel, C., 1997, A&A Suppl. **123**, 241

Fulbright, J.P., Kraft, R.P., 1999, A.J. **118**, 527

Fuhrmann, K., 1998, A&A **338**, 161

Gallagher J.S. III, Hunter, D.A., Tutukov, A.V., 1984, ApJ **284**, 544

Galli, D., Palla, F., Ferrini, F., Penco, U., 1995, ApJ **443**, 536

Garnett, D.R., Shields, G.A., Skillman, E.D., Sagan, S.P., Dufour, R.J., 1997, ApJ **489**, 63

Garnett, D.R., 1998, in "Abundance Profiles: Diagnostic Tools for Galaxy History" ASP Conf. Series, Vol. 147, p. 78

Geisler, D., Friel, E.D., 1992, A.J., **104**, 128

Gerola, H., Seiden, P.E., 1978, ApJ **223**, 129

Gehren, T., Nissen, P.E., Kudritzki, R.P., Butler, K., 1985, in "Production and Distribution of CNO Elements", eds. I.J. Danziger, F. Matteucci and K. Kjaer, (Garching: ESO), 171

Gehrz, R.D., Truran, J.W., Williams, R.E., Starrfield, S., 1998, PASP **110**, 3

Gibson, B.K., 1997, MNRAS **290**, 471

Gilmore, G., Edvardsson, B., Nissen, P.E., 1991 ApJ. **378**, 17

Gilmore, G., Reid, N., 1983, MNRAS **202**, 1025

Gilmore, G., Wyse, R.F.G., 1985, A.J. **90**, 2015

Gilmore G., Wyse, R.F.G., Kuijken, K. 1989 in "Evolutionary Phenomena in Galaxies", eds. J.E. Beckman and B.E.J. Pagel, (Cambridge University Press), p. 172

Gilroy, K.K., Sneden, C., Pilachowski, C.A., Cowan, J.J., 1988, ApJ **327**, 298

Giovagnoli, A., Tosi, M., 1995, MNRAS **273**, 499

Gloecker, G., Geiss, J., 1996, Nature **381**, 210

Gratton, R.G., 1985, A & A, **148**, 105

Gratton, R.G., Sneden, C., 1987, A&A **178**, 179

Gratton, R.G., Carretta, E., Matteucci, F., Sneden, C., 1997a, in "Formation and Evolution of the Galactic Halo: Inside and Out", ASP Conf. Series Vol. 92, p. 307

Gratton, R.G., Carretta, E., Matteucci, F., Sneden, C., 2000, A&A, **358**, 671

Gratton, R.G., Fusi-Pecci, F., Carretta, E., Clementini, G., et al. 1997b, ApJ **491**, 749

Gratton, R.G., Ortolani, S., 1986, A&A **169**, 201

Greggio, L., 1996, in "Interplay between massive star formation, the ISM and galaxy evolution", IAP Meeting July 1995, Editions Frontieres, p. 89

Greggio, L., 1997, MNRAS **285**, 151

Greggio, L., Renzini, A., 1983a, A&A **118**, 217

Greggio, L., Renzini, A. 1983b, in "The First Stellar Generations", Mem. Soc. Astron. It., Vol. 54, p. 311

Grenon, M., 1987, J. Astrophys. Astr. **8**, 123

Grevesse, N., Noels, A., 1993, in "Origin end Evolution of the Elements", eds. N. Prantzos et al. (Cambridge Univ. Press), p. 14

Grevesse, N., Noels, A., Sauval, A.J., 1996, in "Cosmic Abundances", eds. S.S. Holt and G. Sonnenborn, ASP Conf. Series, Vol. 99, p. 117

Guibert, J., Lequeux, J., Viallefond, F., 1978, A&A **68**, 1

Gummersbach, C.A., Kaufer, A., Schaefer, D.R., Szeifert, T., Wolf, B., 1998, A&A **338**, 881

Güsten, R., Mezger, P.G., 1982, Vistas in Astronomy **26**, 159 (GM82)

Hachisu, I., Kato, M., Nomoto, K., 1996, ApJ **470**, L97

Hachisu, I., Kato, M., Nomoto, K., 1999, ApJ **522**, 487

Hartwick, F.D.A., 1976, ApJ **209**, 418

Heger, A., 1998, PhD Thesis, Technische Universität München

Henkel, C., Matthews, H.E., Morris, M., Terebey, S., Fich, M., 1985, A&A **147**, 143

Henry, R.B.C., Pagel, B.E.J., Chincarini, G., 1994, MNRAS **266**, 421

Henry, R.B.C., Howard, J.W., 1995, ApJ **438**, 170

Henry, R.B.C., Pagel, B.E.J., Chincarini, G.L., 1994, MNRAS **266**, 421

Henry, R.B.C., Worthey, G., 1999, PASP **111** 919

Hensler, G. 1988, Habil. Thesis University of Munich (unpublished)

Hensler, G., Burkert, A., 1990, in "Chemical and Dynamical Evolution of Galaxies", eds. F. Ferrini et al., ETS Editrice PISA, p. 168

Hensler, G., Samland, M., Theis, C., Burkert, A., 1993, in "Panchromatic View of Galaxies-their Evolutionary Puzzle", eds. G. Hensler et al., Editions Frontieres, p. 341

Herwig, F. 1996 in "Stellar Evolution: What Should Be Done", 32nd Liege Int. Astrophys. Coll. eds. A. Noels et al., p. 441

Hoyle, F., Fowler, W.A., 1960, ApJ **132**, 565

Iben, I.Jr., 1991, ApJ Suppl. **76**, 55

Iben, I.Jr., Renzini, A., 1983, ARA&A **21**, 271

Iben, I.Jr., Faulkner, J., 1968, ApJ **153**, 101

Iben, I.Jr., Laughlin, G., 1989, ApJ **341**, 312

Iben, I.Jr., Tutukov, A.V., 1984, ApJ Suppl. **54**, 335

Iben, I.Jr., Tutukov, A.V., 1985, ApJ Suppl. **58**, 661

Israelian, G., Garcia-Lopez, R.J., Rebolo, R., 1998, ApJ **507**, 805

Iwamoto, K., Brachwitz, F., Nomoto, K., et al., 1999, ApJ Suppl. **125**, 439

Iwamoto, K., Nakamura, T., Nomoto, K., Mazzali, P.A., Danziger, I.J. et al., 2000, ApJ, **534**, 660

Izotov, Y.I., Chaffee, F.H., Foltz, C.B., Green, R.F., Guseva, N.G., Thuan, T.X., 1999, ApJ **527**, 757

Jacobi, G.H., Ciardullo, R., Ford, H.C., 1990, ApJ **356**, 332

Jeffries, R.D., 1997, MNRAS **288**, 585

Jimenez, R., Flynn, C., Kotoneva, E., 1998, MNRAS **299**, 515

Jones, L.A., Worthey, G., 1995, ApJ Lett. **446**, L31

Josè, J., Hernanz, M., 1998, ApJ **494**, 680

Josey, S.A., Arimoto, N., 1992, A&A **255**, 105

Katz, N., 1992, ApJ **391**, 502

Kaufer, A., Szeifert, T., Krenzin, R., Baschek, B., Wolf, B., 1994, A&A **289**, 740

Kauffmann, G., White, S.D.M., Guiderdoni, B., 1993, MNRAS **264**, 201

Kennicutt, R.C. Jr., 1983, ApJ **272**, 54

Kennicutt, R.C. Jr., 1989, ApJ **344**, 685

Kennicutt, R.C. Jr., 1998a, ARA&A **36**, 189

Kennicutt, R.C. Jr., 1998b, ApJ **498**, 541

Kennicutt, R.C. Jr., Garnett, D.R., 1996, ApJ **456**, 504

Kennicutt, R.C. Jr., Tamblyn, P., Congdon, C.E., 1994, ApJ **435**, 22

Kerr, F.J., 1969, ARA&A **7**, 39

Kilian-Montenbruck, J., Gehren, T., Nissen, P.E., 1994, A&A **291**, 757

Kirshner R.P., 1990 in "Supernovae" ed. A. Petschek, A.A.L. Springer-Verlag, p. 59

Köppen, J., Theis, Ch., Hensler, G., 1995, A&A **296**, 99

Kormendy, J., Djorgovski, S., 1989, ARA&A **27**, 235

Kroupa, P., Tout, C.A., Gilmore, G., 1993, MNRAS **262**, 545

Kuijken, K., Gilmore, G., 1989, I MNRAS **239**, 571

Kuijken, K., Gilmore, G., 1989, II MNRAS **239**, 605

Kuijken, K., Gilmore, G., 1989, III MNRAS **239**, 651

Kuijken, K., Gilmore, G., 1991, ApJ **367**, L9

Kulkarni, S.R., Heiles, C., Blitz, L., 1982, ApJ **259**, 63

Kunth, D., Matteucci, F., Marconi, G., 1995, A&A **297**, 634

Lacey, C.G., Fall, S.M., 1985, ApJ **290**, 154

Laird, J.B., 1985, ApJ **289**, 556

Landau, L.D., Lifshitz, E.M., 1962, "Quantum Mechanics" (London: Pergamon)

Langer, N., Henkel, C., 1995, Space Sci. Rev. **74**, 343 (LH95)

Larson, R.B., 1969, MNRAS **145**, 405

Larson, R.B., 1972, Nature Phys. Sci. **236**, 7

Larson, R.B., 1974, MNRAS **169**, 229

Larson, R.B., 1975, MNRAS **173**, 671

Larson, R.B., 1976, MNRAS **176**, 31

Larson, R.B., 1986, MNRAS **218**, 409

276

Larson, R.B., 1991 in "Frontiers of Stellar Evolution" ed. D.L. Lambert, ASP Conf. Series Vol. 20, p. 571

Larson, R.B., 1998, MNRAS **301**, 569

Lawler, J.E., Whaling, W., Grevesse, N., 1990, Nature, **346**, 635

Lee, Y.W., 1992, A.J. **104**, 1780

Lee, Y.W., Demarque, P., Zinn, R., 1994, ApJ **423**, 248

Leitherer, C., Robert, C., Drissen, L., 1992, ApJ **401**, 596

Li, Ti Pei, Riley, P.A., Wolfendale, A.W., 1982, J. Phys. G. **8**, 1141

Liller, W., Mayer, B., 1987, PASP **99**, 606

Lin, D.N.C., Pringle, J.E., 1987, ApJ **320**, L87

Linsky, J.L., Brown, A., Gayley, K. et al., 1994, ApJ **402**, 694

Linsky, J.L., Diplos, A., Wood, B.E. et al., 1995, ApJ **451**, 335

Livio, M., 2000, in "Supernovae and Gamma-Ray Bursts: the greatest explosions since the Big Bang", ed. M. Livio, N. Panagia and K. Sahu, Cambridge Univ. Press, p. 334

Liu, X.-W., Storey, P.J., Barlow, M.J., Danziger, I.J., Cohen, M., Bryce, M. 2001, MNRAS, **312**, 585

Luck, R.E., Bond, H.E., 1985, ApJ **292**, 559

Luminet, J.P., in "Black Holes", Cambridge Univ. Press

Lynden-Bell, D., 1975, Vistas in Astronomy **19**, 299

Lynden-Bell, D., 1977, IAU Symp. Vol. 75, ed. T. de Jong and A. Maeder, p. 291

Lyne, A.G., Manchester, R.N., Taylor, J.H., 1985, MNRAS **213**, 613

MacFadyen, A.I., Woosley, S.E., 1999, ApJ **524**, 262

Mac Low, M.-M., Ferrara, A., 1999, ApJ **319**, 471

Maciel, W.J., Quireza, C., 1999, A&A **345**, 629

Maciel, W.J., Köppen, J., 1994, A&A **282**, 436

Maciel, W.J., Chiappini, C., 1994, ApSS **219**, 231

Madau, P., della Valle, M., Panagia, N., 1998, MNRAS **297**, L17

Maeder, A., 1981, A&A **102**, 401

Maeder, A., 1983, A&A **120**, 113

Maeder, A., 1992, A&A **264**, 105 (M92)

Maeder, A., 1993, A&A **268**, 833

Maeder, A., 1999, in IAU Symp. Vol. 193, ed. K. van der Hucht et al., p. 177

Maeder, A., Meynet, G., 1989, A&A **210**, 155

Maeder, A., Meynet, G., 1987, A&A **182**, 243

Malaney, R.A., Mathews, G.J., Dearborn, D.S.P., 1989, ApJ **345**, 169

Malinie, G., Hartmann, D.H., Clayton, D.D., Mathews, G.J., 1993, ApJ **413**, 633

Majewsky, S.R., 1993, ARA&A, **31**, 575

Marconi, G., Matteucci, F., Tosi, M., 1994, MNRAS **270**, 35

Marigo, P., 2000, in "The Chemical Evolution of the Milky Way: Stars versus Clusters", eds. F. Matteucci and F. Giovannelli, Kluwer (Dordrecht), ASSL Vol. 255, p. 481 (M2K)

Marigo, P., Bressan, A., Chiosi, C., 1996, A&A **313**, 545

Marquez, A., Schuster, W.J., 1994, A&A Suppl. Series **108**, 341

Martin, P., Roy, J.R., 1994, ApJ **424**, 599

Martinelli, A., Matteucci, F., Colafrancesco, S., 1998, MNRAS **298**, 42

Mathews, W.G., Baker, J.C., 1971, ApJ **170**, 241

Mathews, G.J., Alcock, C.R., Fuller, G.M., 1990, ApJ **349**, 449

Mathews, G.J., Schramm, D.N., 1988, ApJ Lett. **324**, L67

Matteucci, F., 1986, MNRAS **221**, 911

Matteucci, F., 1989, in "Evolutionary Phenomena in Galaxies", eds. J.E. Beckman and B.E.J. Pagel, Cambridge Univ. Press, p. 297

Matteucci, F., 1991a, in "Frontiers of Stellar Evolution" ed. D.L. Lambert, ASP Conf. Series, Vol. 20, p. 539

Matteucci, F., 1991b, in "SN1987A and Other Supernovae", eds. I.J. Danziger and K. Kjär, ESO Publ., p. 703

Matteucci, F., 1994, A&A **288**, 57

Matteucci, F., Brocato, E., 1990, ApJ **365**, 539

Matteucci, F., Chiosi, C., 1983, A&A **123**, 121

Matteucci, F., Chiappini, C., 1999, in "Chemical Evolution from Zero to High Redshift", ESO Astrophysics Symposia, eds. J.R. Walsh and M.R. Rosa, p. 83

Matteucci, F., D'Antona, F., Timmes, F.X., 1995, A&A **303**, 460

Matteucci, F., Franco, J., François, P., Treyer, M.A., 1989, Rev. Mex. Astron. Astrophys. **18**, 145

Matteucci, F., François, P., 1992, A&A **262**, L1

Matteucci, F., François, P., 1989, MNRAS **239**, 885

Matteucci, F., Gibson, B.K., 1995, A&A **304**, 11

Matteucci, F., Greggio, L., 1986, A&A **154**, 279

Matteucci, F., Molaro, P., Vladilo, G., 1997, A&A **321**, 45

Matteucci, F., Padovani, P., 1993, ApJ **419**, 485

Matteucci, F., Raiteri, C., Busso, M., Gallino, R., Gratton, R., 1993, A&A **272**, 421

Matteucci, F., Romano, D., Molaro, P., 1999, A&A **341**, 458

Matteucci, F., Tornambè, A., 1985, A&A **142**, 13

Matteucci, F., Tornambè, A., 1987, A&A **185**, 51

Matteucci, F., Vettolani, G., 1988, A&A **202**, 21

Mayle, R., Wilson, J.R., 1988, ApJ **334**, 909

Mayor, M., Vigroux, L., 1981, A&A **98**, 1

Mazzali, P.A., Lucy, L.B., Danziger I.J., et al., 1993, A&A **269**, 423

Mazzitelli, I., D'Antona, F., Caloi, V., 1995, A&A **302**, 382

McCall M.L., Rybski, P.M., Shields, G.A., 1985, ApJ Suppl. **57**, 1

McCall M.L., 1982 PhD Thesis, University of Texas, Austin

McClure, R.D., VandenBerg, D.A., Smith, G.H., Fahlman, G.G., Richer, H.B., Hesser, J.E., Harris, W.E., Stetson, P.B., Bell, R.A., 1986, ApJ **307**, L49

McWilliam, A., 1997, ARA&A, **35**, 503

McWilliam, A., Rich, R.M., 1994, ApJ Suppl. **91**, 749

McWilliam, A., Preston, G.W., Sneden, C., Searle, L., 1995, A.J. **109**, 2736

Melendez, J., Barbuy, B., 2002, ApJ **575**, 474

Merrifield, M., 1992, A.J. **103**, 1552

Meusinger, H., Stecklum, B., Reimann, H.-G., 1991, A&A **245**, 57

Meyer, B.S., Schramm, D.N., 1986, ApJ **311**, 406

Michard, R., 1983, A&A **121**, 313

Miller, G.E., Scalo, J.M., 1979, ApJ Suppl. **41**, 513

Minniti, D., 1993, Ph.D. Thesis, Steward Observatory, University of Arizona

Minniti, D., Olszewski, E.W., Liebert, J., White, S.D.M., Hill, J.M., Irwin, M.J., 1995 MNRAS **277**, 1293

Minniti, D., Vandehei, T., Cook, K.H., Griest, K., Alcock, C. 1998, ApJ **499**, L175

Mirabel, I.F., Morras, R., 1984, ApJ **279**, 86

Mirabel, I.F., Morras, R., 1990, ApJ **356**, 130

Mirabel, I.F., 1989, in IAU Coll. 120 "Structure and Dynamics of the Interstellar Medium", eds. M. Moles et al., Springer-Verlag, p. 396

Mitchell, R.J., Culhane, J.L., Davison, P.J.N., Ives, J.C., 1976 MNRAS, **175**, 29

Mo, H.J., Mao, S., White, S.D.M., 1998, MNRAS **295**, 319

Molaro, P., Bonifacio, P., Castelli, F., Pasquini, L., 1997, A&A **319**, 593

Molla, M., Ferrini, F., 1995, ApJ **454**, 726

Molla, M., Ferrini, F., Diaz, A.I., 1996, ApJ **466**, 668

Monteverde, M.I., Herrero, A., Lennon, D.J., Kudritzki, R.P., 1997, ApJ **474**, L107

Ng, Y.K., Bertelli, G., 1998, A&A **329**, 943

Nissen, P.E., Schuster, W.J., 1997, A&A **326**, 751

Nomoto, K., 1984, ApJ **277**, 791

Nomoto, K., 1987, ApJ **322**, 206

Nomoto, K., Iwamoto, K., Suzuki, T., 1995, Physics Reports **256**, 173

Nomoto, K., Thielemann, F.K., Yokoi, K., 1984, ApJ **286**, 644

Norris, J.E., 1986, ApJ Suppl. **61**, 667

Norris, J.E., Ryan, S.G., Stringfellow, G.S., 1994, ApJ **423**, 386

Oey, M.S., Kennicutt, R.C. Jr., 1993, ApJ **411**, 137

Olive, K.A., 1999, in "Theoretical and Observational Cosmology Summer School", Cargese, Corsica, NATO Science Series, Vol. 541, p. 36

Olive, K.A., Steigman, G., Skillman, E.D., 1997, ApJ, **483**, 788

Olive, K.A., Fields, B.D., 1999 in ASP Conf. Series, ed. R. Ramaty et al., Vol. 171, p. 36

Oort, J.H., 1970, A&A **7**, 381

Opik, E.J., 1953, Irish Astron. J., Vol. 2(8), p. 219

Ortolani, S., Renzini, A., Gilmozzi, R., et al., 1995, Nature **377**, 701

Padovani, P., Matteucci, F., 1993, ApJ **416**, 26

Pagel, B.E.J., 1989, in "Evolutionary Phenomena in Galaxies", eds. J.E. Beckman and B.E.J. Pagel, Cambridge Univ. Press, p. 201

Pagel, B.E.J., 1994, V IAC Winter School in "Galaxy Formation and Evolution", eds. C. Munoz-Tunon and F. Sanchez, Cambridge Univ. Press

Pagel, B.E.J., 1997, in "Nucleosynthesis and Chemical evolution of Galaxies", Cambridge Univ. Press

Pagel, B.E.J., Patchett, B.E., 1975, MNRAS **172**, 13

Pagel, B.E.J., Portinari, L., 1998, MNRAS **298**, 747

Pagel, B.E.J., Simonson, E.A., Terlevich, R.J., Edmunds, M.G., 1992, MNRAS **255**, 325

Pain, R., Hook, I.M., et al., 1996, ApJ, **473**, 356

Pardi, M.C., Ferrini, F., 1994, ApJ, **421**, 491

Pardi, M.C., Ferrini, F., Matteucci, F., 1995, ApJ **444**, 207

Pasquali, A., Perinotto, M., 1993, A&A **280**, 581

Peebles, D.J.E., Dicke, R.H., 1968, ApJ **154**, 891

Peimbert, M., 1978, IAU Symp. 76, ed. Y. Terzian, Reidel-Dordrecht, p. 215

Peimbert, M., 1993, Rev. Mex. Astron. Astrophys. **27**, 9

Perlmutter, S., Aldering, G., Goldhaber, G. et al., 1999, ApJ **517**, 565

Persic, M., Salucci, P., Stel, F., 1996, MNRAS **281**, 27

Peterson, R.C., Kurucz, R.L., Carney, B.W., 1990, Apj, **350**, 173

Pettini, M., Smith, L.J., King, D.L., Hunstead, R.W., 1997, ApJ **486**, 665

Pettini, M., Ellison, S.L., Steidel, C.C., Bowen, D.V., 1999, ApJ **510**, 576

Pilyugin, L.S., 1993, A&A **277**, 42

Plez, B., Smith, V.V., Lambert, D.L., 1993, ApJ **418**, 812

Pont, F., Mayor, M., Turon, C., Vandenberg, D.A., 1998, A&A **329**, 87

Portinari, L., 1998, PhD Thesis, University of Padua

Portinari, L., Chiosi, C., Bressan, A., 1998, A&A **334**, 505

Prantzos, N., Boissier, S., 2000, MNRAS, **313**, 338

Prantzos, N., Vangioni-Flam, E., Chauveau, S., 1994, A&A **285**, 132

Prantzos, N., Casse, M., Vangioni-Flam, E., 1993, ApJ **403**, 630

Prantzos, N., Aubert, O., Audouze, J., 1996, A&A **309**, 760

Prantzos, N., Aubert, O., 1995, A&A **302**, 69

Prantzos, N., 1996, A&A **310**, 106

Primas, F., Molaro, P., Castelli, F., 1994, A&A **290**, 885

Raiteri, C., Villata, M., Navarro, J.F., 1996, A&A **315**, 105

Rana, N.C., 1991, ARA&A **29**, 129

Rana, N.C., Basu, S., 1990, ApSS **168**, 317

Rana, N.C., Wilkinson, D.A., 1986, MNRAS **218**, 497

280

Ratnatunga, K.U., van den Bergh, S., 1989, ApJ **343**, 713

Rebolo, R., Beckman, J.E., Molaro, P., 1988, A&A **192**, 192

Reeves, H., 1993, A&A **269**, 166

Renzini, A., 1994, A&A **285**, L5

Renzini, A., Voli, M., 1981, A&A **94**, 175 (RV81)

Rich, R.M., 1988, A.J. **95**, 828

Rich, R.M., 1990, ApJ **362**, 604

Rocha-Pinto, H.J., Maciel, W.J., 1996, MNRAS, **279**, 447

Rocha-Pinto, H.J., Maciel, W.J., Scalo, J., Flynn, C., 1999, ApSS **265**, 245

Romano, D., Matteucci, F., Molaro, P., Bonifacio, P., 1999, A&A **352**, 117

Rosenberg, A., Saviane, I., Piotto, G., Aparicio, A., 1999, A.J., **118**, 2306

Roy, J.R., Kunth, D., 1995, A&A **294**, 432

Rudolph, A.L., Simpson, J.P., Haas, M.R., Erickson, E.F., Fich, M., 1997, ApJ **489**, 94

Russel, S.C., Bessel, M.S., Dopita, M.A., 1988, in "The Impact of High S/N Spectroscopy on Stellar Physics" eds. G. Cayrel de Strobel and M. Spite, Reidel (Dordrecht), p. 545

Ryan, S.G., Norris, J.E., Bessel, M.S., 1991, A.J. **102**, 303

Ryan, S.G., Norris, J.E., Beers, T.C., 1996, ApJ **471**, 254

Ryan, S.G., Norris, J.E., 1991, A.J. **101**, 1865

Ryan, S.G. 2000, in "The Galactic Halo: from Globular Clusters to Field Stars", 35th Liege Int. Astrophys. Coll., p. 101

Ryder, S., 1998, in "Abundance Profiles: Diagnostic Tools for Galactic History", ASP Conf. Series, Vol. 147, p. 83

Sackmann, I.J., Boothroyd, A.I., 1999, ApJ **510**, 217

Sadler, E.M., Rich, R.M., Terndrup, D.M., 1996, A.J. **112**, 171

Sadoulet, B., 1999, Rev. Mod. Phys. **71**, 197

Salpeter, E.E., 1955, ApJ **121**, 161

Sandage, A., 1987, in "The Galaxy", eds. G. Gilmore and B. Carswell, (Reidel, Dordrecht), p. 321

Sandage, A., 1988, ApJ **331**, 583

Sandage, W.J., Fouts, G., 1987, A.J. **93**, 74

Sanders, D.B., Solomon, P.M., Scoville, N.Z., 1984, ApJ **276**, 182

Scalo, J.M., 1986, Fund. Cosmic Phys. **11**, 1

Scalo, J.M., 1998, in "The Stellar Initial Mass Function", ASP Conf. Series, Vol. 142, p. 201

Schaller, G., Schaerer, D., Meynet, G., Maeder, A., 1992, A&A Suppl. **96**, 296 (SSMM92)

Schmidt, M., 1959, ApJ **129**, 243

Schmidt, M., 1963, ApJ **137**, 758

Schuster, W.J., Nissen, P.E., 1989, A&A **222**, 69

Searle, L., Zinn, R., 1978, ApJ **225**, 357 (SZ)

Serrano, A., 1986, PASP **98**, 1066

Sharov, A.S., 1972, SvA **16**, 41

Shaver, P.A., McGee, R.X., Newton, L.M., Danks, A.C., Pottasch, S.R., 1983, MNRAS **204**, 53

Shore, S.N., Ferrini, F., 1995, Fund. Cosmic Phys. **16**, 1

Shore, S.N., Ferrini, F., Palla, F., 1987, ApJ **316**, 663

Schuster, W.J., Nissen, P.E., 1989 A&A **222**, 69

Scully, S., Casse, M., Olive, K.A. et al., 1996, ApJ **462**, 960

Skillman, E.D., Kennicutt, R.C.Jr., 1993, ApJ **411**, 655

Skillman, E.D., Kennicutt, R.C.Jr., Shields, G.A., Zaritsky, D. 1996 Ap.J. **462**, 147

Silk, J., 1980, in "Star Formation" 10th Advanced Course Swiss Society of Astronomy and Astrophysics Saas-Fe, eds. A. Maeder and L. Martinet, p. 133

Sikivie, P., 1983, Phys. Rev. Lett. **51**, 1415

Simpson, J.P., Colgan, S.W.J., Rubin, R.H., Erickson, E.F., Haas, M.R., 1995, ApJ **444**, 721

Smartt, S.J., Rolleston, W.R.J., 1997, ApJ Lett. **481**, L47

Smith, V.V., Lambert, D.L., 1989, ApJ **345**, L75

Smith, V.V., Lambert, D.L., 1990, ApJ Lett. **361**, L69

Smith, V.V., Lambert, D.L., Nissen, P.E., 1993, ApJ **408**, 262

Sneden, C., Crocker, D.A., 1988, ApJ **335**, 406

Sneden, C., Gratton, R.G., Crocker, D.A., 1991, A&A **246**, 354

Sneden, C., Lambert, D.L., Whitaker, R.W., 1979, ApJ **234**, 684

Sommer-Larsen, J., 1991, MNRAS **249**, 368

Sommer-Larsen, J., Yoshii, Y., 1989, MNRAS **238**, 133

Songaila, A., Cowie, L.L., Weaver, H., 1988, ApJ **329**, 580

Spite, M., Spite, F., 1978, A&A **67**, 23

Spite, M., Spite, F., 1990, A&A **234**, 67

Stark, A.A., 1984, ApJ **281**, 624

Starrfield, S., Truran, J.W., Sparks, W.M., Arnould, M., 1978, ApJ **222**, 600

Steinmetz, M., Müller, E., 1995, MNRAS **276**, 549

Steinmetz, M., Müller, E., 1994, A&A **281**, L97

Strobel, A., 1991, A&A **247**, 35

Talbot, R.J., 1981, SPIE **264**, 291

Talbot, R.J., Arnett, D.W., 1971, ApJ **170**, 409 (TA71)

Talbot, R.J., Arnett, D.W., 1973, ApJ **186**, 69

Talbot, R.J., Arnett, D.W., 1975, ApJ **195**, 551

Tammann, G.A., Loeffler, W., Schröder, A., 1994, ApJ Suppl. **92**, 487

Tammann, G.A., Leibungut, B., 1990, A&A **230**, 81

Tammann, G.A., Schröder, A., 1990 A&A **236**, 149

Tayler, R.J., 1990, Q.Jl RAS **31**, 281

Tenorio-Tagle, G., 1980, A&A **88**, 61

Terndrup, D.M., 1988, A.J. **96**, 884

Theis, Ch., Burkert, A., Hensler, G., 1992, A&A **265**, 465

Thielemann, F.K., Nomoto, K., Hashimoto, M., 1996, ApJ **460**, 408 (TNH95 and TNH96)

Thielemann, F.K., Nomoto, K., Hashimoto, M., 1993, in "Origin and Evolution of the Elements", ed. N. Prantzos et al., Cambridge Univ. Press p. 297

Thielemann, F.K., Hashimoto, M.A., Nomoto, K., 1990, ApJ **349**, 222

Thielemann, F.K., Metzinger, J., Klapdor, H.V., 1983, A&A **123**, 162

Thomas, D., Schramm, D.N., Olive, K.A., Fields, B.D., 1993, ApJ **406**, 569

Thornburn, J.A., 1994, ApJ **421**, 318

Timmes, F.X., Woosley, S.E., Weaver, T.A., 1995, ApJ Suppl. **98**, 617

Tinsley, B.M., 1980, Fund. Cosmic Phys. **5**, 287

Tomkin, J., Woolf, V.M., Lambert, D.L., 1995, A.J. **109**, 2204

Tomkin, J., Sneden, C., Lambert, D.L., 1986, ApJ **415**, 420

Toomre, A., 1964, ApJ **139**, 121

Tornambè, A., Chieffi, A., 1986, MNRAS **220**, 529

Tornambè, A., Matteucci, F., 1987, ApJ Lett. **318**, L25

Tosi, M., 1996, in "From Stars to Galaxies: the Impact of Stellar Physics on Galaxy Evolution", ed. C. Leitherer et al., ASP Conf. Series, Vol. 98 p. 299

Tosi, M., 1982, ApJ **254**, 699

Tosi, M., 1988a, A&A **197**, 33

Tosi, M., 1988b, A&A **197**, 47

Tosi, M., Diaz, A.I., 1990, MNRAS **246**, 616

Tosi, M., Steigman, G., Matteucci, F., Chiappini, C., 1998, ApJ **498**, 226

Travaglio, C., Galli, D., Gallino, R., Busso, M., Ferrini, F. Straniero, O., 1999a, ApJ **521**, 691

Travaglio, C., Galli, D., Gallino, R., Busso, M., Ferrini, F., Straniero, O., 1999b, in "Nuclei in the Cosmos V" ed. N. Prantzos, Editions Frontieres p. 531

Tsujimoto, T., Yoshii, Y., Nomoto, K., Shigeyama, T., 1995, A&A **302**, 704

Twarog, B.A., 1980, ApJ **242**, 242

Twarog, B.A., Ashman, K.M., Anthony-Twarog, B.J., 1997, A.J. **114**, 2556

Vanbeveren, D., Vanrensbergen, W., De Loore, C., 1994, in "The Brightest Binaries", Kluwer Academic Publ.

Vandenberg, D.A., Bolte, M., Stetson, P.B., 1990, A.J. **100**, 445

van den Bergh, S., 1988, Comments on Astrophysics **12**, 131

van den Bergh, S., 1991, ApJ **369**, 1

van den Bergh, S., McClure, R.D., 1989, ApJ **347**, L29

van den Bergh, S., Tammann, G., 1991, ARA&A **29**, 363

van den Bergh, S., 1962, A.J. **62**, 486

van den Bergh, S., 1995, ApJ **450**, 27

van den Hoek, L.B., Groenewegen, M.A.T., 1997, A&A Suppl. **123**, 305(HG97) van der Hulst, J.M., Skillman, E.D., Kennicutt, R.C., Bothun, G.D., 1987, A&A **177**, 63

van der Marel, R.P., 1991, MNRAS **253**, 710

van der Marel, R.P., Binney, J., Davies, R.L., 1990, MNRAS **245**, 582

van Zee, L., 1998, in "Abundance Profiles: Diagnostic Tools for Galaxy History", ed. D. Friedli et al., ASP Conf. Series, Vol. 147, p. 98

Vangioni-Flam, E., Audouze, J., 1988, A&A **193**, 81

Venn, K.A., McCarthy, J.K., Lennon, D.J., Kudritzki, R.P., 1998, in "Abundance Profiles: Diagnostic Tools for Galaxy History", ASP Conf. Series, Vol. 47, p. 54

Vila-Costas, M.B., Edmunds, M.G., 1992, MNRAS **259**, 121

Vila-Costas, M.B., Edmunds, M.G., 1993, MNRAS **265**, 199

Vilchez, J.M., Esteban, C., 1996, MNRAS **280**, 720

Walker, T., Viola, V.E., Mathews, G.J., 1985, ApJ **299**, 745

Wallerstein, G., Iben, I. Jr., Parker, P. et al., 1997, Rev. Mod. Phys. **69**, 995

Wallerstein, G., Greenstein, J.L., Parker, R., Helfer, H.L., Aller, L.H., 1963, ApJ **137**, 280

Warmels, R.H., 1988, A&A Suppl. **73**, 453

Weaver, T.A., Woosley, S.E., 1993, Physics Reports **227**, 65

Webb, J.K., Carswell, R.F., Lanzetta, K.M. et al., 1997, Nature **388**, 250

Weidemann, V., 1987, A&A **188**, 74

Wheeler, J.C., Sneden, C., Truran, J.W., 1989, ARA&A **27**, 279

Whelan, J., Iben, I. Jr., 1973, ApJ **186**, 1007

Wielen, R., 1977, A&A **60**, 263

Wielen, R., Fuchs, B., Dettbarn, C., 1996, A&A **314**, 438

Wilkinson, M., Evans, N., 1999, MNRAS **310**, 645

Wilson, T.L., Matteucci, F., 1992, Astron. Astrophys. Rev. **4**,1

Wilson, T.L., Rood, R.T., 1994, ARA&A **32**, 191

Wood, M.A., 1990, J.R.Astron.Soc.Can. **84**, 150

Woosley, S.E., Hartmann, D.H., Hoffman, R.D., Haxton, W.C., 1990, ApJ **356**, 272

Woosley, S.E., Axelrod, T.S., Weaver, T.A., 1984, in "Stellar Nucleosynthesis", eds. C. Chiosi and A. Renzini, Reidel (Dordrecht) p. 263

Woosley, S.E., Weaver, T.A., 1986, ARA&A **24**, 205

Woosley, S.E., 1987, in "Nucleosynthesis and Chemical Evolution", 16th Saas Fee Course, eds. B. Hauck et al. (Geneva: Geneva Observatory), p. 1

Woosley, S.E., Langer, N., Weaver, T.A., 1993, ApJ **411**, 823

Woosley, S.E., Langer, N., Weaver, T.A., 1995, ApJ **448**, 315

Woosley, S.E., Weaver, T.A., 1988, Physics Reports **163**, 79

Woosley, S.E., Weaver, T.A., 1994, ApJ **423**, 371

Woosley, S.E., Weaver, T.A., 1995, ApJ Suppl. **101**, 181(WW95)

Wyse, R.F.G., Gilmore, G., 1993, in "The Globular Cluster-Galaxy Connection" eds. J. Brodie and G. Smith., ASP Conf. Series, Vol. 48, p. 727

284

Wyse, R.F.G., Gilmore, G., 1995, A.J. **110**, 2771
Wyse, R.F.G., Gilmore, G., 1992, A.J. **104**, 144
Wyse, R.F.G., Silk, J., 1987, ApJ **313**, L11
Wyse, R.F.G., Silk, J., 1989, ApJ **339**, 700
Yokoi, K., Takahashi, K., Arnould, M., 1983, A&A **117**, 65
Yoshii, Y., 1982, PASJ **34**, 365
Yoshii, Y., Saio, H., 1979, PASJ **31**, 339
Yungelson, L., Livio, M., 1998, ApJ **497**, 168
Yungelson, L., Livio, M., Tutukov, A., 1997, ApJ **481**, 127
Zaritsky, D., Kennicut, R.C., Huchra, J.P., 1994, ApJ **420**, 87
Zinn, R., 1980, ApJ **241**, 602
Zinn, R., 1985, ApJ **293**, 424
Zinn, R., 1993, in " The Globular Cluster-Galaxy Connection", eds. J. Brodie
and G. Smith, ASP Conf. Series, Vol. 48, p. 38

Acknowledgments

This book could not have been completed without the help of several friends and colleagues. First of all I thank Cristina Chiappini for the long discussions about the chemical evolution of the Galaxy during the period in which she was preparing her PhD thesis here in Trieste, for reading some the chapters and for patiently providing some of the figures of this book. Thanks to Donatella Romano and Massimo Frutti for creating with great care some of the tables and the figures appearing in the book. Then my deep gratitude goes to my husband, John Danziger, who patiently read many chapters of the book. I am also indebted to several colleagues who read at least one chapter of the book and provided suggestions: Monica Tosi, Rob Kennicutt, Friedel Thielemann, Renzo Sancisi, Agostino Martinelli, Nicolai Chugai, Simone Recchi and Giovanni Carraro. I also thank all the colleagues who gave permission of reproducing figures from their papers and in particular those who provided also the files of the original figures, these are: Max Pettini, Annette Ferguson, Dick Henry, Pier Carlo Bonifacio, Stuart Ryder, Walter Maciel, Paola Marigo, Liese van Zee, Claudia Travaglio, Thomas Dame, Laura Portinari, Donatella Romano and Monica Tosi.

Special thanks to Butler Burton who proposed me to write such a book and to all the friends, relatives and colleagues who encouraged me to start and complete this enterprise.